Necessity of
Project Engineering
Knowledge
in Engineering Curriculum
Subhendu Moulik
MW00973208

AuthorHouse™ UK Ltd.
500 Avebury Boulevard
Central Milton Keynes, MK9 2BE
www.authorhouse.co.uk
Phone: 08001974150

First published by AuthorHouse 4/15/2011

ISBN: 978-1-4567-7927-6 (sc)

This book is printed on acid-free paper.

This book is dedicated to:

my wife Lopamudra, son Abhinav and Aaron

– *Subhendu Moulik*

Acknowledgements

I am grateful to
Mr. Sinu Mathew Zachariah, Mr. Varghese Abraham,
Mr. Frank Feely, Mr. Gregory J. Fox, Mr. Michael Fletcher
and
Mr. Abrie van Blommestein
for their suggestions and support
in composing this book.

Project Implementation Review

Project Definition

Project Communication

Project Closure

Project Planning

Monitoring & Control

Detailed Planning

Project Execution

Introduction

Normally, engineers graduate from engineering schools and universities as single discipline engineers. After graduation, they will take up a job based on their particular engineering discipline and then progress in their career. Engineering, Procurement, Construction (EPC) projects are very much the norm in the present day engineering contracting industry for capital projects of all types and sizes. In the EPC projects environment all engineering disciplines works together as a team. Every individual discipline has its own role together with the role of exchanging information with other disciplines at the right time. There is no doubt that graduates from engineering schools and colleges have learnt their discipline well. However they have very sketchy knowledge about other discipline design deliverables and will have almost no knowledge about inter dependency of design information. Many universities conduct summer projects where different disciplines students get involved to complete a common task. This effort is indeed appreciable. However if Universities think that by conducting a summer project they have finished their duty towards a multi discipline and practical engineering education then they are incorrect. These days industry requires that students know much more than the regular conventional engineering design studies. Universities guide students with a conventional study method, where value engineering, cost optimization, quality engineering, construction , safety, operability engineering aspects may be given less importance or the university has no means to teach students these subjects.

Students entering into the real world working environment find that design information and drawings for example are in unfamiliar formats. They need help from colleagues and superiors to understand what they feel they should know. They begin their working lives feeling confused, helpless and vulnerable and this is where the struggle begins.

The project engineering study as a subject will close these loopholes in the standard engineering curriculum. Universities, engineering schools and colleges may find this concept new or strange. It is logical and necessary to teach students project engineering to add more value to their career. In the following chapters the subject of project engineering is defined together with examples. Effort has been made to give faculty members and students an impression of what the minimum requirements are for teaching Project Engineering as a subject.

Contents

Chapter 1

What is Project Engineering?

Project Engineering is nothing but the integration of all the different aspects of an EPC project.

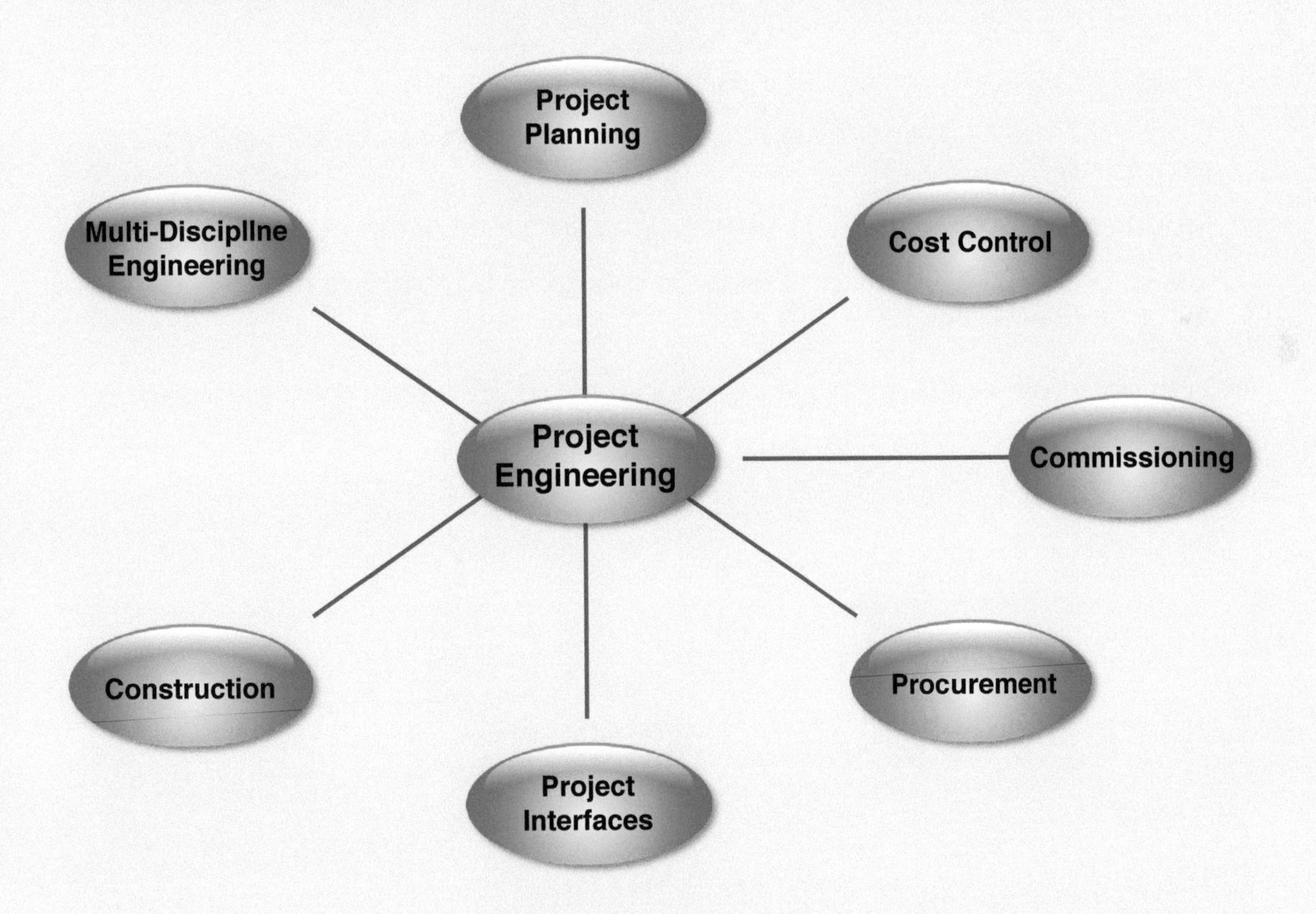

Project Engineering is composed of the following subjects, that engineering students to know as a minimum.

1. Multi discipline project engineering – KITCS rule
2. Project planning – Schedule the activities of a project
3. Cost Control – Engineering cost estimation, know the cost components of a project and know how to derive it
4. Project Interfaces – What are they and how to manage it?
5. Construction – Implementation of design work to reality – what are the basic principles?
6. Procurement – Is it something like shopping in a mall or different?
7. Risk analysis and mitigation – What are the risk of a project and how to mitigate them?

In later chapters, each of above components' are described and analysed, why a student to know about it.

Where is Project Engineering Applicable?

Any design that needs multiple discipline engineering input and confirmation.

Example:
Design of a factory (Nuclear, Chemical, Petrochemical, Refinery, Oil & Gas etc.)

Engineering discipline involved:

1. Chemical Engineering
2. Piping Engineering
3. Civil Engineering
4. Electrical Engineering
5. Instrumentation Engineering
6. Mechanical Static and Rotary
7. Technical Health & Safety

Project engineering knowledge will ensure that each discipline fulfills other discipline's requirements and finally achieve the need of the project.

Example of engineering discipline's inter dependability during engineering design

Process Engineering	Piping Engineering
1. As a support document to design basis, process engineering provides Process flow diagram to piping discipline	2. Based on information's received from process discipline, piping engineers makes preliminary layout study and suggest location of new equipment and location of pipe rack if new pipe rack becomes essential.
3. Process discipline issues hand mark-up P&ID to piping identifying new equipment(s), line size changes from existing scheme and any other changes	4. Piping releases isometrics for process hydraulics. Forward bulk materials take off for procurement.
5. Process issues tie in P&ID and tie in schedule identifying type of tie in type e.g. during shut down during local shutdown or hot tapping during plant operation .	6. Piping reviews feasibility of tapping and give feedback to management.

Why Project Engineering?

This can be explained as below.

NO Project Engineering Knowledge!! What Happens next?

1. No integration check during design
2. Designers unaware about potential clash
3. Designer unaware that design objective will not be fulfilled
4. Designer is unaware that design is UNSAFE
5. Chaos during implementation/construction
6. Rework
7. Delay in Project
8. Budget Over-run
9. Accident/Breakdown of machinery during operation due to lack of design integrity

Picture of Clash

Example of Clash between
Pipe and Steel Structure

Example of Clash between
Cable Tray and Steel Structure

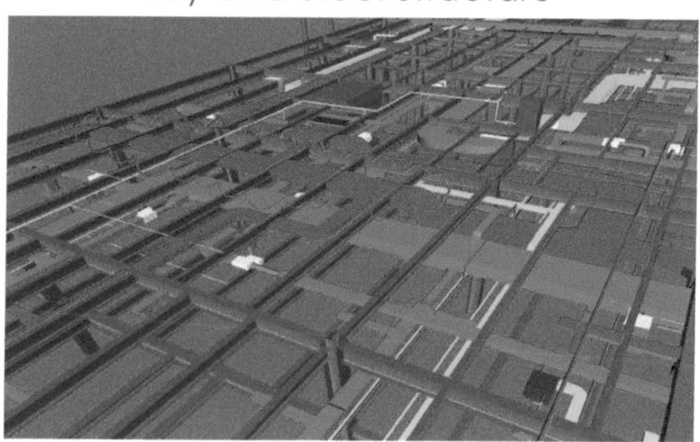

Example of Clash between
Plumbing and Duct

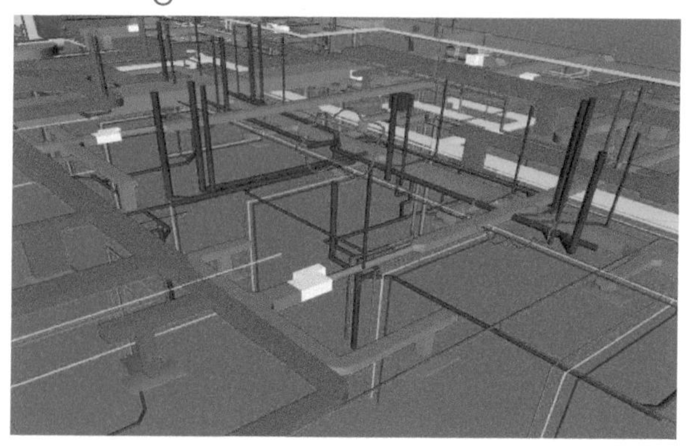

Example of Clash between
Piping and Underground Cable Trench

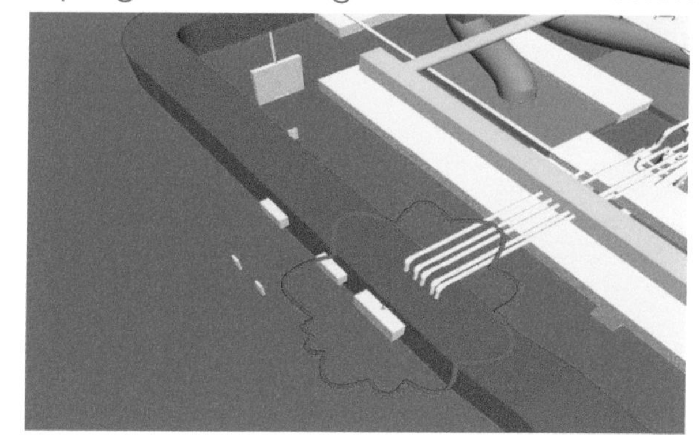

Why is Project Engineering knowledge necessary for students?

Project engineering is engineering co-ordination between all or specific engineering disciplines, construction, procurement Project control and contracts. In a multiple discipline engineering environment, engineering disciplines are generally dependant on other discipline(s), either to provide input or review other discipline's document. Project engineer/Engineering management ensure proper review of all discipline deliverables.

This is like a car, where all components are functioning properly, however it is not moving. Multi-disciple project engineers ensure this car moves and runs efficiently project engineering knowledge will make design flawless, safe and fulfill design objectives. It avoids rework, budget and schedule overrun.

Project engineering provides the needed link between industry and education system and public welfare.

University Graduates to have the following basic knowledge when entering into industry

- How different engineering disciplines are interrelated
- What are the major engineering deliverables of all disciplines
- What these deliverables look like
- Basic idea of man-hours or time required to produce an engineering document
- To learn what is multi discipline project engineering and how to function efficiently in a multi discipline working environment
- To be aware of current industry requirements
- To be aware of various commissioning and construction related problems which could be solved during design stages
- To learn about maintainability and constructability
- To learn about engineering standard cost and schedule
- To learn how to do design right in the very first place
- To be aware of design safety features and benefit
- It will make the student confident
- Students will have more credentials and will be more attractive to industry
- Can contribute immediately while entering industry and will help an individual through out his/her carrier path

How Students will be benefited from Project Engineering?

Normally engineers graduate from engineering schools and universities as single discipline engineers. After graduation, they will take up a job based on their particular engineering discipline and then progress in their career. Project Engineers combine the different disciplines into a frame work which is often referred to as multi-discipline Project Engineering.

- To learn what is multi-discipline project engineering and how to function efficiently in a multi-discipline working environment
- To be aware of current industry requirements
- To be aware of various commissioning and construction related problems which could be solved during design stages
- To learn about maintainability and constructability
- To learn about engineering standard cost and schedule
- To learn how to do design right at the very first place
- To be aware of design safety features and benefits
- It will make students confident
- Students will have more credentials and will be more attractive to industry
- Can contribute immediately while entering industry and will help an individual throughout his/her career path

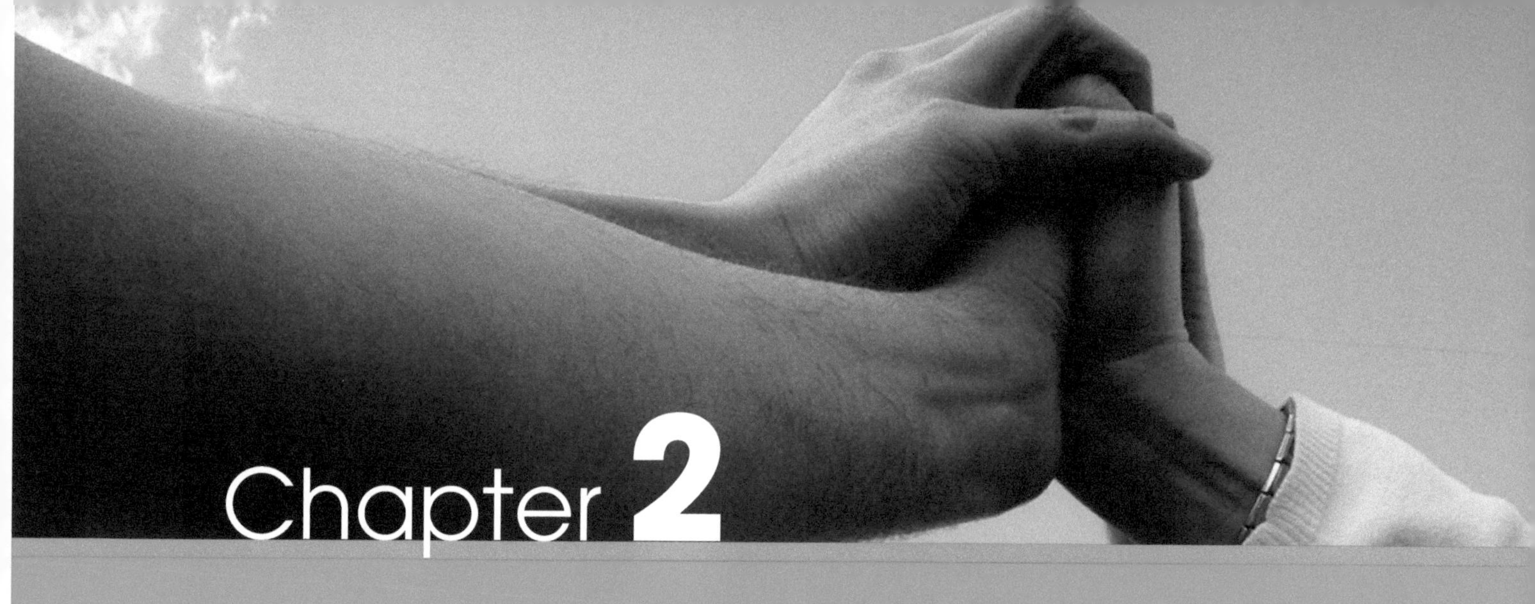

Chapter 2

Multi-disciplinary Project Engineering & KITCS Rule

Multi-discipline project engineering is a special skill where the engineer has to have basic knowledge of all discipline deliverables.

We have the following basic engineering disciplines in EPC industry.

1. Chemical engineering – We often call it Process engineering
2. Mechanical engineering – It is very vast subject and divided into three disciplines.
   a. Mechanical Static – Work with Static vessel design and fabrication
   b. Mechanical Rotary – Work with rotary mechanical items, e.g. Pump, compressor
   c. Mechanical Piping – Often called as piping engineer, who works on piping layout, flexibility, materials design as well as implementation.
3. Electrical engineering – Works on electrical design
4. Instrumentation engineering – Has again two disciplines
   a. Instrument discipline – work on Instrumentation hardware
   b. Process control discipline – Work on software part of instrumentation
5. Civil engineering – Works on Civil design and construction.

Project Engineers/Engineering Managers combine the different disciplines into a frame work which is often referred to as multi-discipline Project Engineering. Multi- discipline Project Engineers/Managers require in-depth knowledge of how different disciplines are interrelated so that they can ensure proper review and flow of work between different disciplines before any document is issued for implementation.

Multi-discipline project engineering is a special skill where the engineer has to have basic knowledge of all discipline deliverables. One must know the criticality of the document and how long it should take to get reviewed/commented by other discipline. Multi-discipline project engineers should have understanding of Quality, be cost consciousness and have an understanding of how deliverables effects scheduling. It is an added bonus if one has knowledge of risk management.

Multi-discipline project engineering is a subject by itself, until unless it is taught, students will never know about it.

Discipline wise % engineering man hour distribution of a typical project is described below.

Discipline wise man-hour distribution

Discipline	% Man-hours
Process Engineering	10
Static Equipment Rotating Equipment Package Equipment	15
Electrical Engineering	10
Instrumentation Engineering	15
Piping layout Piping Design Piping Stress and Support	30
Civil & Structural Engineering	20

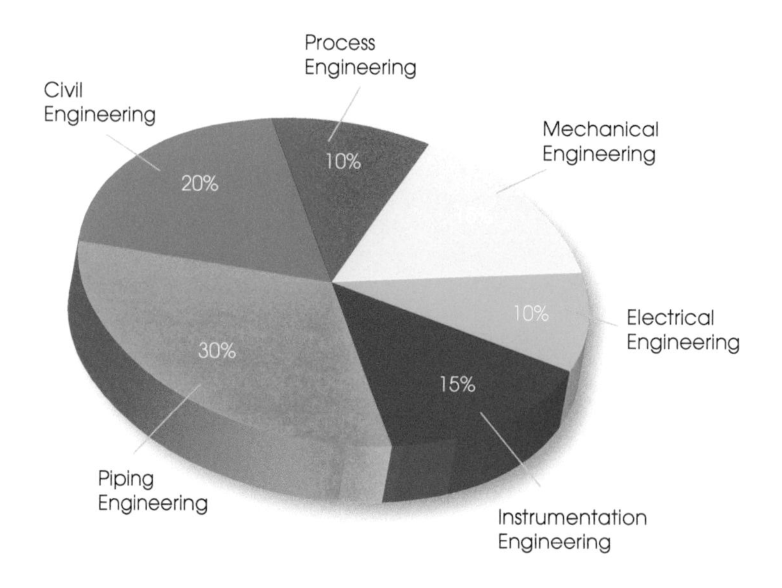

Below are examples, how inter-discipline co-ordinations takes place during a multi-disciplinary engineering project.

In an EPC Project within the Oil & Gas industry, Process Engineers play a key role through-out the project life cycle. Chemical engineers, specialists in plant simulation, initiate a project by providing heat and mass balance reports for the project. Process engineers keep contributing in the project by developing further detailed engineering drawings, e.g. piping & instrumentation diagrams, engineering data sheets etc. The Process group leads hazard studies (HAZOP) through-out the engineering phase.

The Piping discipline begins its activity with the preparation of a conceptual Plot Plan for the project. Layout of major equipment items, interconnecting pipe racks, flares, storage tanks, utility units are considered during basic engineering Layout. The layout of the facility goes through a multi-discipline review to consider safety and operational aspects of the plant. The disciplines which form part of the layout are Process, Piping, HSE, Commissioning, Civil etc. Piping engineering continues its journey in a project typically through piping layout, piping modelling, 3D model review, isometrics, preparation of material take-offs and requisitions.

Electrical and Instrumentation personnel start working with finalising location of major items e.g. Substations, Field auxiliary room, main control room etc. They also do play a significant role in project cost estimation before client takes investment decision. During detail engineering Electrical and Instrumentation initially get input from Piping engineering on plot plan, equipment location, control valve location etc. As the project progresses they proceed with their own deliverables. Electrical and instrumentation disciplines do work for each other. Instrumentation panel's electric supply is designed by Electrical where as motor trip signal is managed by Instrumentation Engineer.

Civil is the popular discipline that works in conjunction with Piping, Electrical and Instrumentation. It designs the foundations for all disciplines. Within the Civil group the detailed design for heavy duty (HD) paving is undertaken, the areas of the Plant requiring HD paving is ascertained from the Piping Layout group and is shown on the site wide Plot Plan. The Civil group also design trenches and in many consultancies, the Civil group take responsibility for underground sewerage drainage. Civil's also look after the design for surface drainage by way of 'water run-off' paving.

Process Control discipline works very closely with Process Engineering and Instrumentation. This discipline take up plant operation control logic from operation philosophy, cause & effect diagrams and ultimately implement process and operation philosophy in to real life plant operation.

The multi-disciplinary project engineering is (KITCS Rule)

- **K** Knowledge of all engineering disciplines and their standard engineering design deliverables
- **I** Inter dependency rule – how one engineering discipline design deliverable is dependant on other discipline's input and confirmation.
- **T** Team work
- **C** Cost consciousness
- **S** Schedule and risk consciousness

"Teamwork is the ability to work together toward a common vision.

The ability to direct individual accomplishments toward organizational objectives.

It is the fuel that allows common people to attain uncommon results."

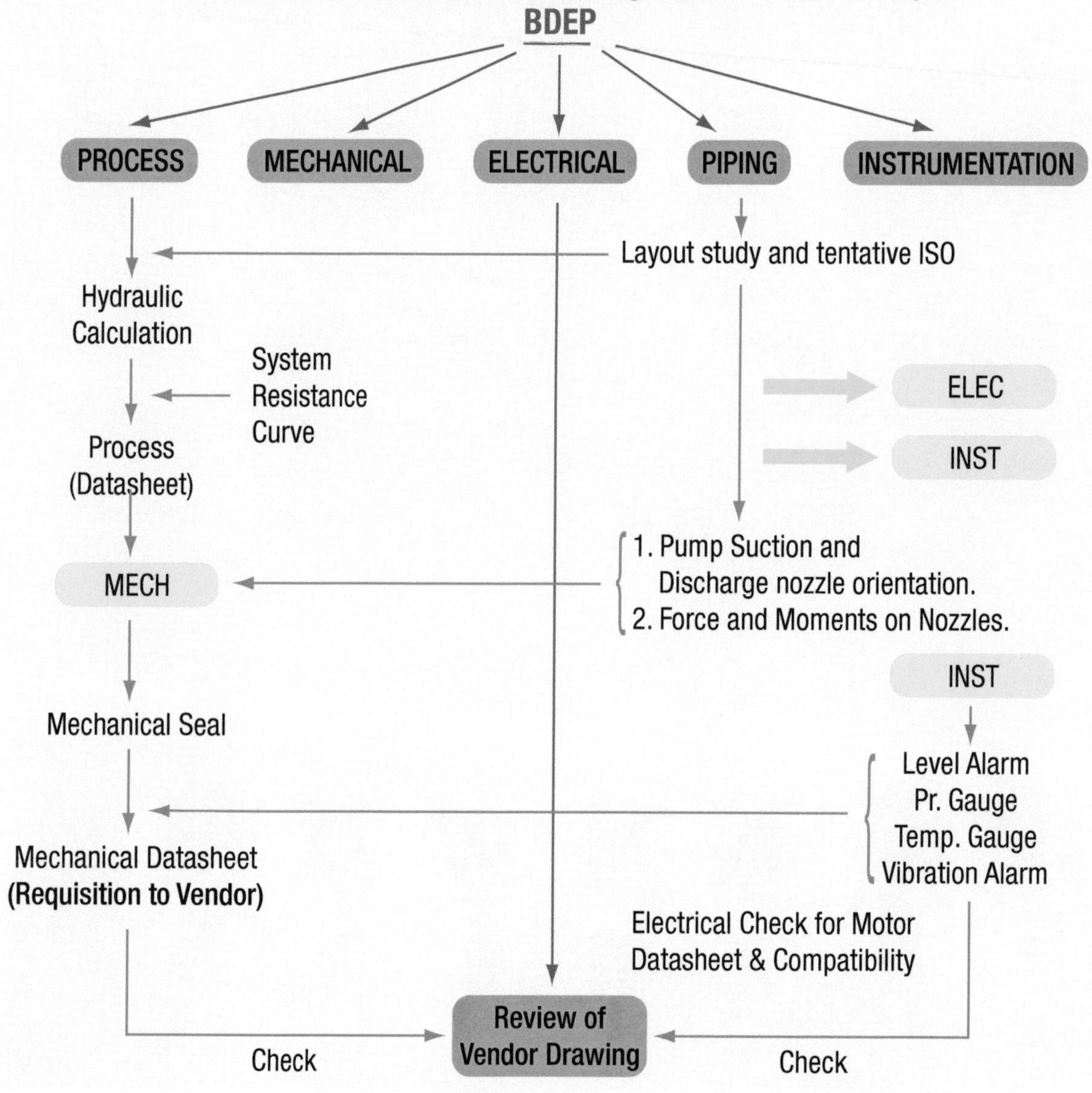
Example of Multi-discipline Engineering Work Flow
Addition of Pump in existing system (Revamp)
BDEP
PROCESS
MECHANICAL
ELECTRICAL
PIPING
INSTRUMENTATION
Layout study and tentative ISO
Hydraulic Calculation
System Resistance Curve
ELEC
INST
Process (Datasheet)
1. Pump Suction and Discharge nozzle orientation.
2. Force and Moments on Nozzles.
MECH
INST
Mechanical Seal
Level Alarm
Pr. Gauge
Temp. Gauge
Vibration Alarm
Mechanical Datasheet
(Requisition to Vendor)
Electrical Check for Motor Datasheet & Compatibility
Review of Vendor Drawing
Check
Check

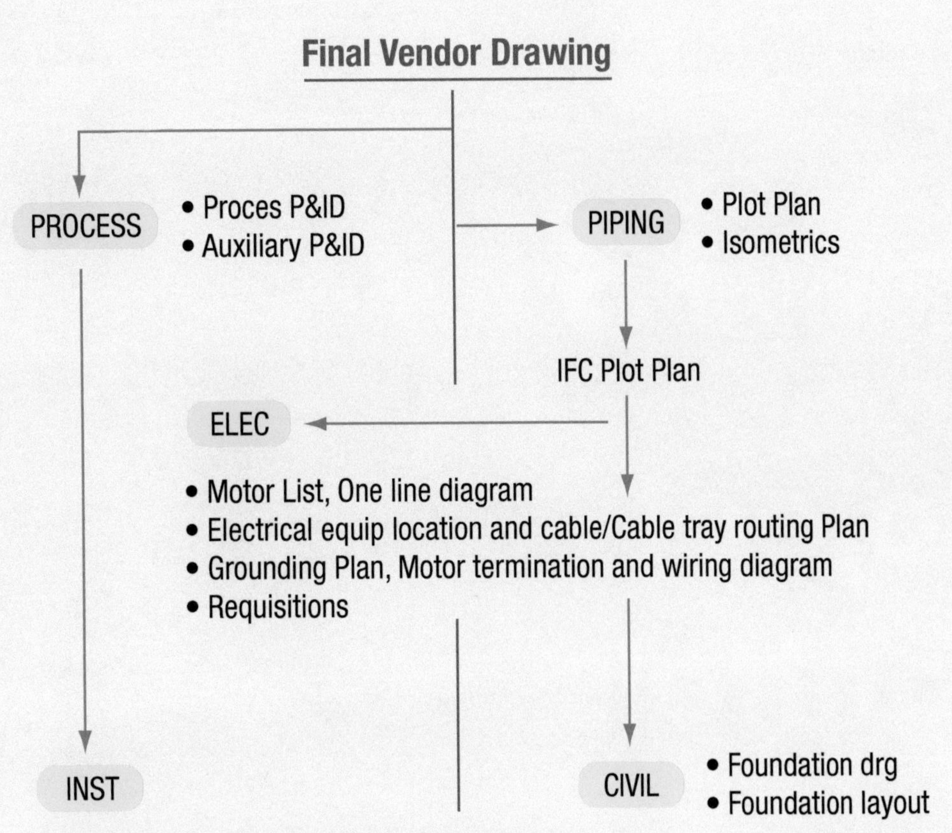

• Inst. index, Cable layout. Inst. location plan
• Set point and alarm summary list, DCS configuration
• Requisitions

Construction

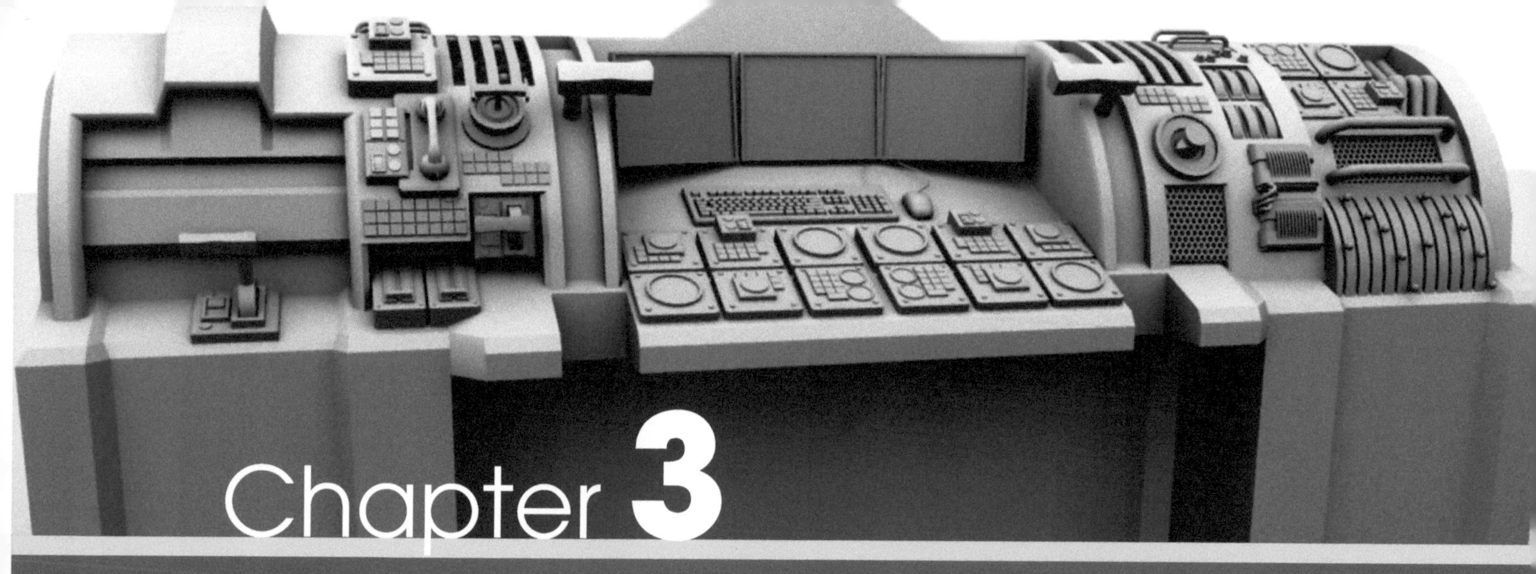

Chapter 3

Project Controls

What is it?

Project Controls is a form of project management that involves the overseeing of projects and assignments goals and outcomes for a department or entire company. Project Controls makes sure that a project is completed on time and within budget. Project Control Manager leads project controls team.

What are the roles of a Project Control Manager?

1. Act as liaison between partners, associates, subcontractors and construction sites to make sure that their own schedules are coherent with the general schedule.
2. Establish along with relevant Managers, the required resources for relevant parts of the project, including staffing etc to ensure that these are accurate and realistic and that any necessary steps needed to supply resource have been arranged with reference to the critical path.
3. Reviews or assists in reviewing proposal provisions related to schedule/cost engineering and material control, and develops supporting data for contract negotiations.
4. Identifies cost trends and schedule impacts for management attention.
5. Member of risk committee and custodian of risk register.
6. Establish effective monitoring and reporting systems to ensure that Company's Policy is consistently implemented.
7. Review and assess the integrity, frequency & value of all reporting in respective area, establishing revised systems where necessary to ensure that Project management have the necessary up to date information.
8. Monitor equipment, materials, buildings & other assets during demobilization. Ensuring that all punch-list items are resolved, client hand over is satisfactorily concluded.

The organization chart of project controls department is as below.

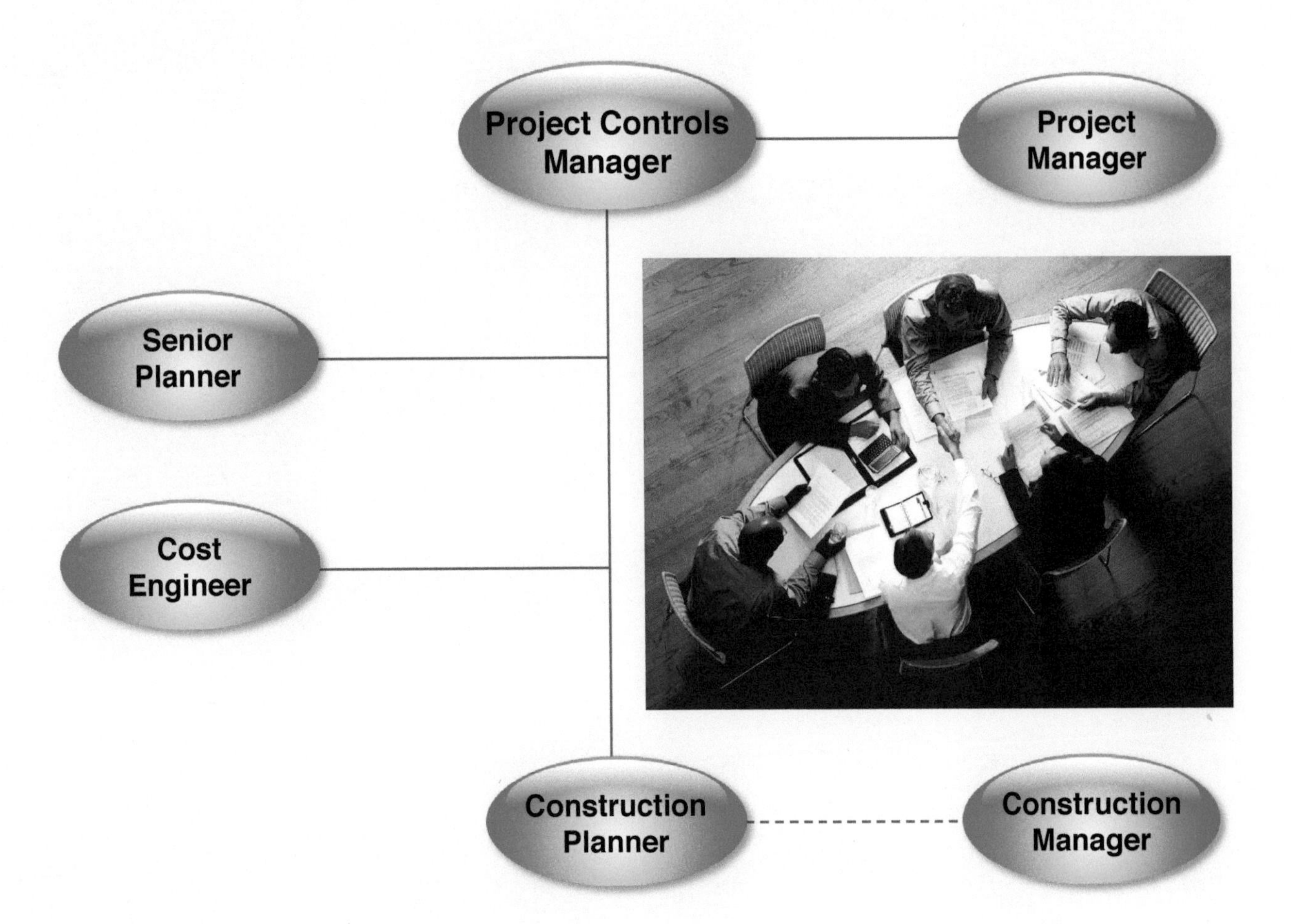

Chapter 4

Project Planning

Project Planning has to identify and breakdown the whole project work into small activities in a logical manner. This is called the Work Breakdown Structure or WBS. The WBS gives a visual representation of the order and relationship of activities. It is necessary to develop the WBS to avoid missing any element of project activity. With the WBS, reviewers of the schedule find it easy to review and able to identify mistakes if any. Once the WBS is ready, it is the time to assign activity duration and prepare your project schedule.

The project planning does the following basic works

1. Preparation of WBS
2. Preparation of Project Responsibility Matrix
3. Assignment of responsible person or group for each activity
4. Assignment of duration to each and every activity
5. Assignment of Cost to every activity
6. Assigning of resource to every activity
7. Relationship with one activity to the other
8. Forecast project duration
9. Manage Critical path

Every engineering discipline has it's own panning department who co-ordinates with Project central planning group.

Why students are to learn Project planning?

1. Will learn how to identify each and every activity of a project work
2. Will have an impression of duration of activities related to engineering, construction, pre commissioning and commissioning.

WORK BREAKDOWN STRUCTURE (WBS) – ENGINEERING

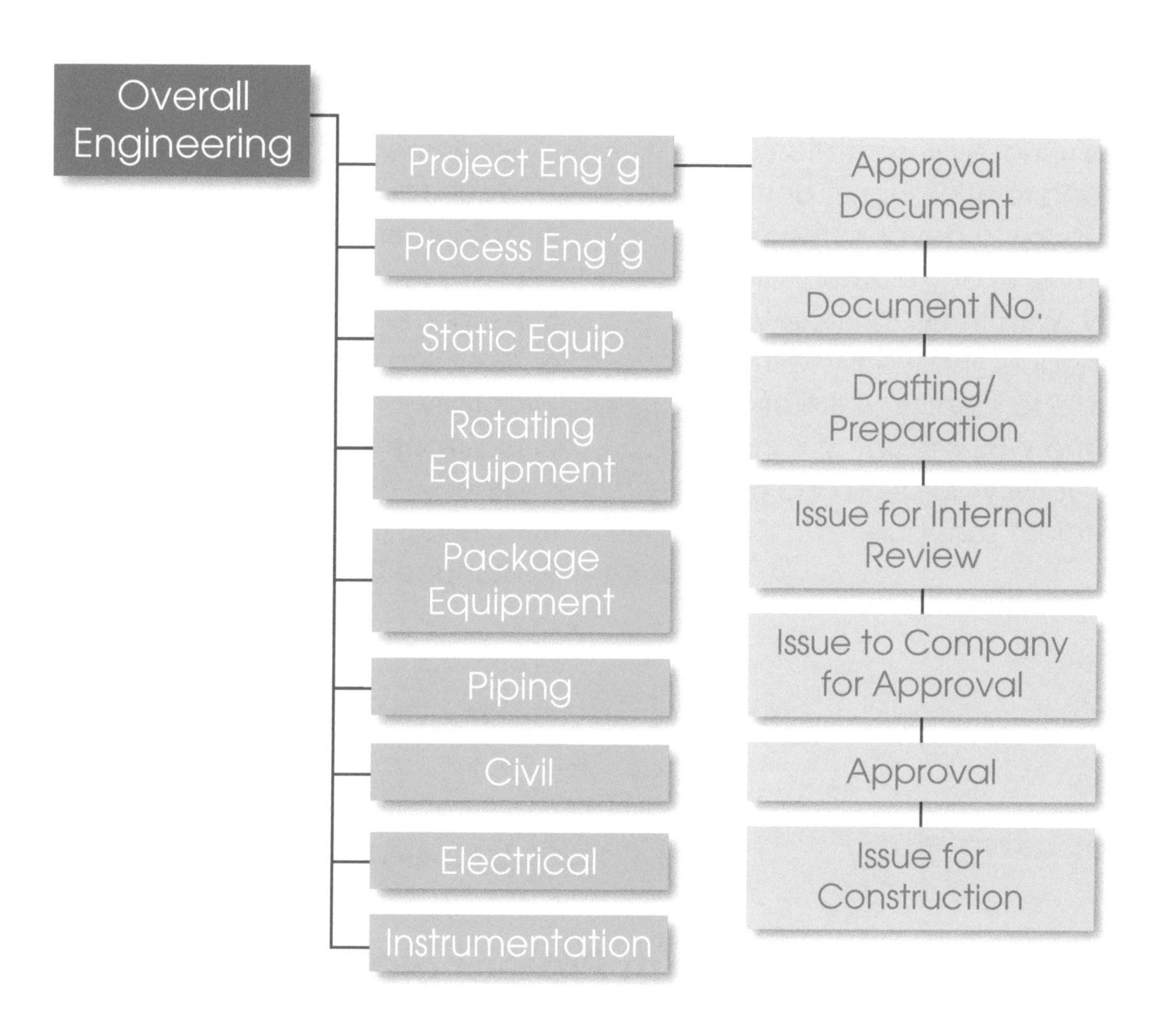

WORK BREAKDOWN STRUCTURE (WBS) – PROCUREMENT

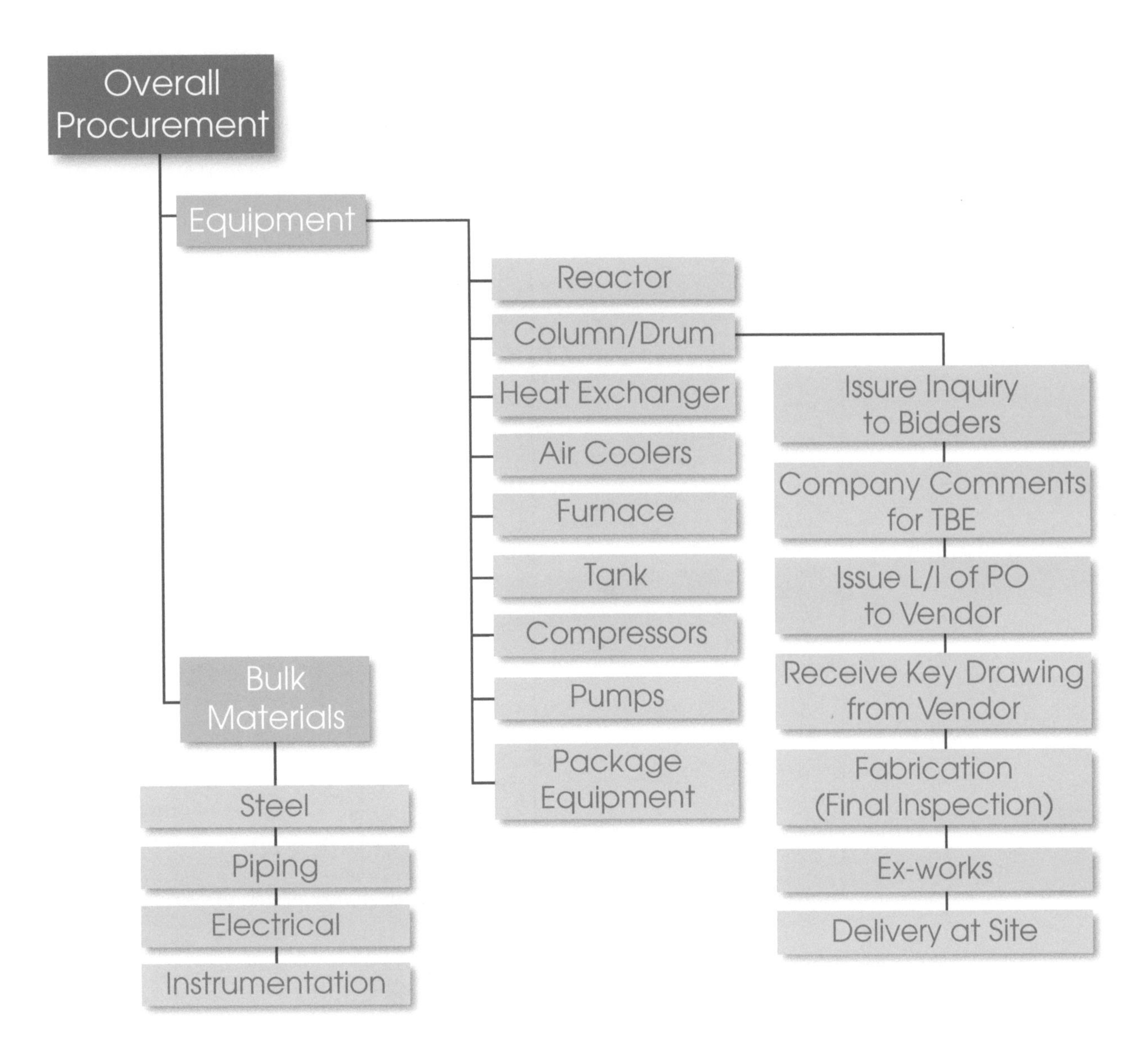

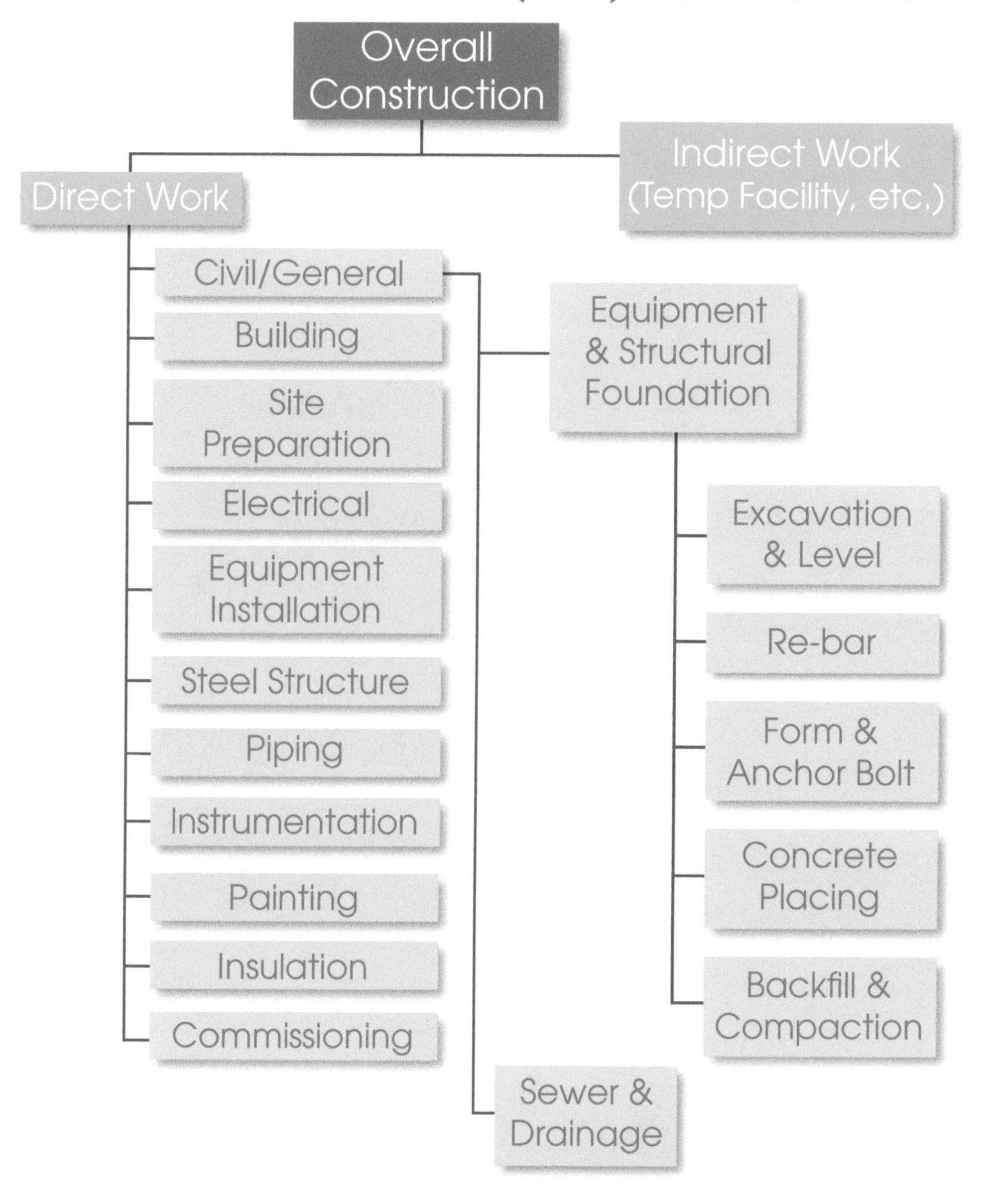
WORK BREAKDOWN STRUCTURE (WBS) – CONSTRUCTION
Overall Construction
Direct Work
Indirect Work (Temp Facility, etc.)
Civil/General
Building
Site Preparation
Electrical
Equipment Installation
Steel Structure
Piping
Instrumentation
Painting
Insulation
Commissioning
Equipment & Structural Foundation
Excavation & Level
Re-bar
Form & Anchor Bolt
Concrete Placing
Backfill & Compaction
Sewer & Drainage

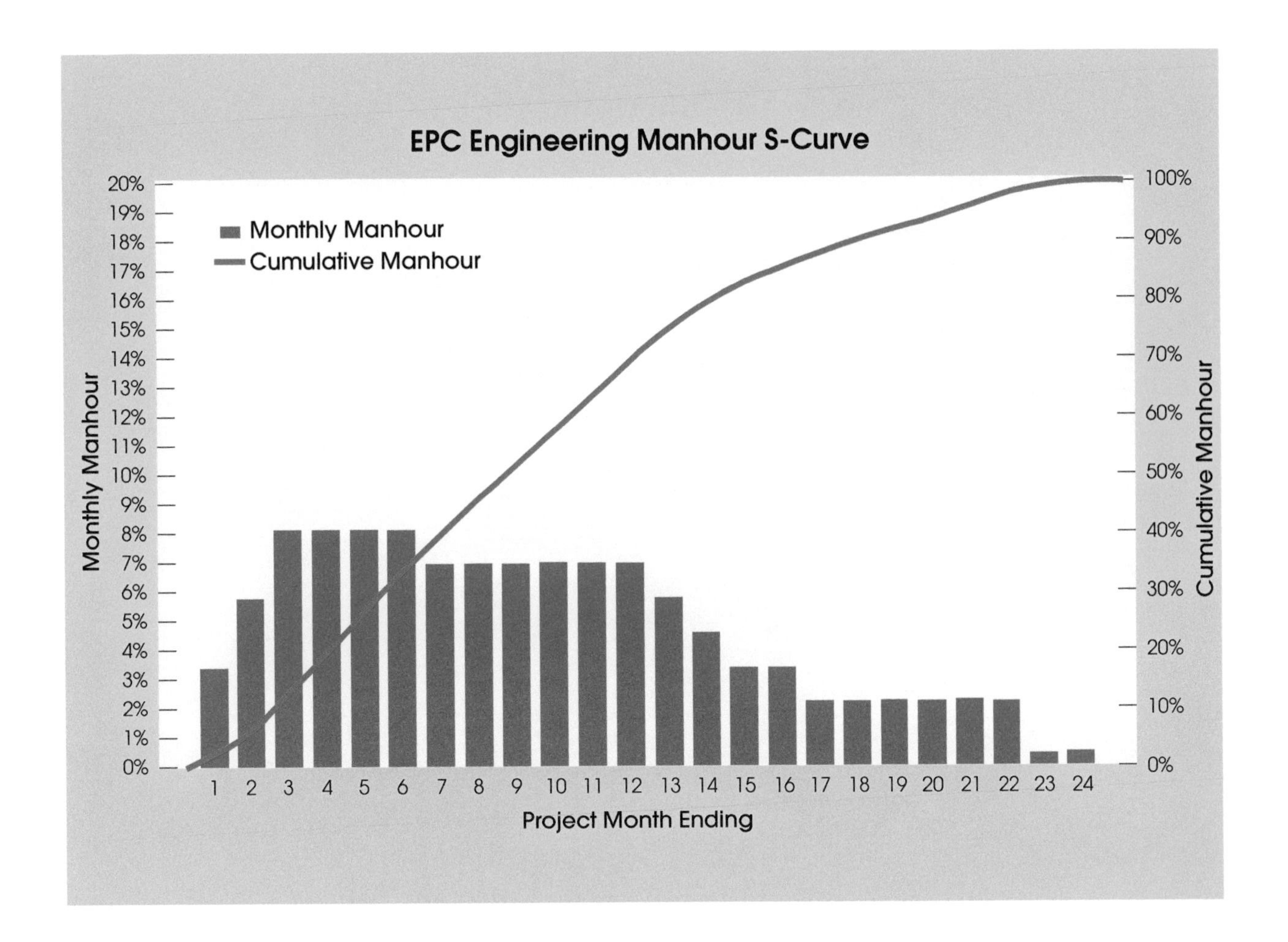
EPC Engineering Manhour S-Curve
Monthly Manhour
Cumulative Manhour
Monthly Manhour
Cumulative Manhour
Project Month Ending
20%
19%
18%
17%
16%
15%
14%
13%
12%
11%
10%
9%
8%
7%
6%
5%
4%
3%
2%
1%
0%
100%
90%
80%
70%
60%
50%
40%
30%
20%
10%
0%
1 2 3 4 5 6 7 8 9 10 11 12 13 14 15 16 17 18 19 20 21 22 23 24

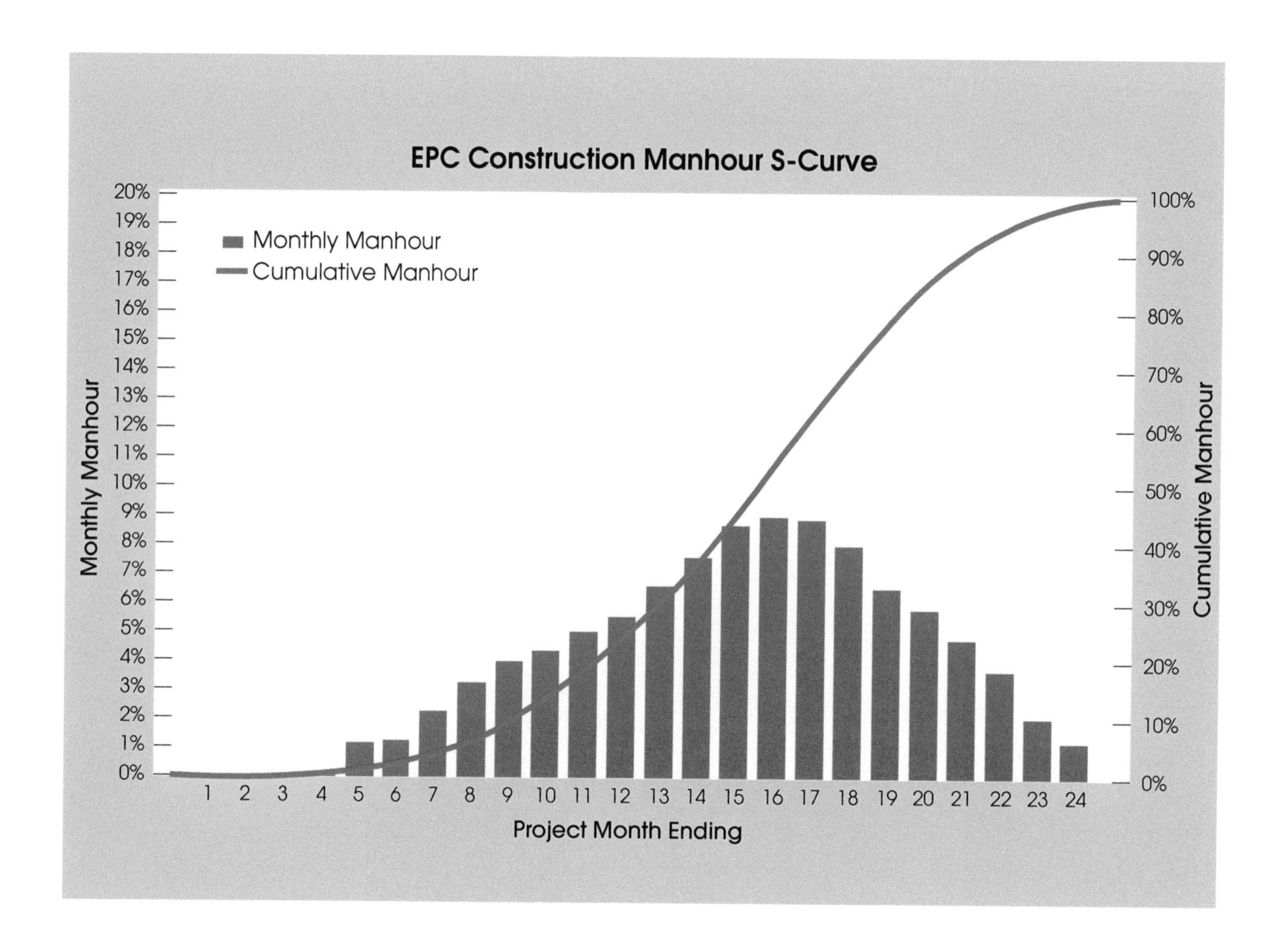
EPC Construction Manhour S-Curve
Monthly Manhour
Cumulative Manhour
Monthly Manhour
20%
19%
18%
17%
16%
15%
14%
13%
12%
11%
10%
9%
8%
7%
6%
5%
4%
3%
2%
1%
0%
Cumulative Manhour
100%
90%
80%
70%
60%
50%
40%
30%
20%
10%
0%
1 2 3 4 5 6 7 8 9 10 11 12 13 14 15 16 17 18 19 20 21 22 23 24
Project Month Ending

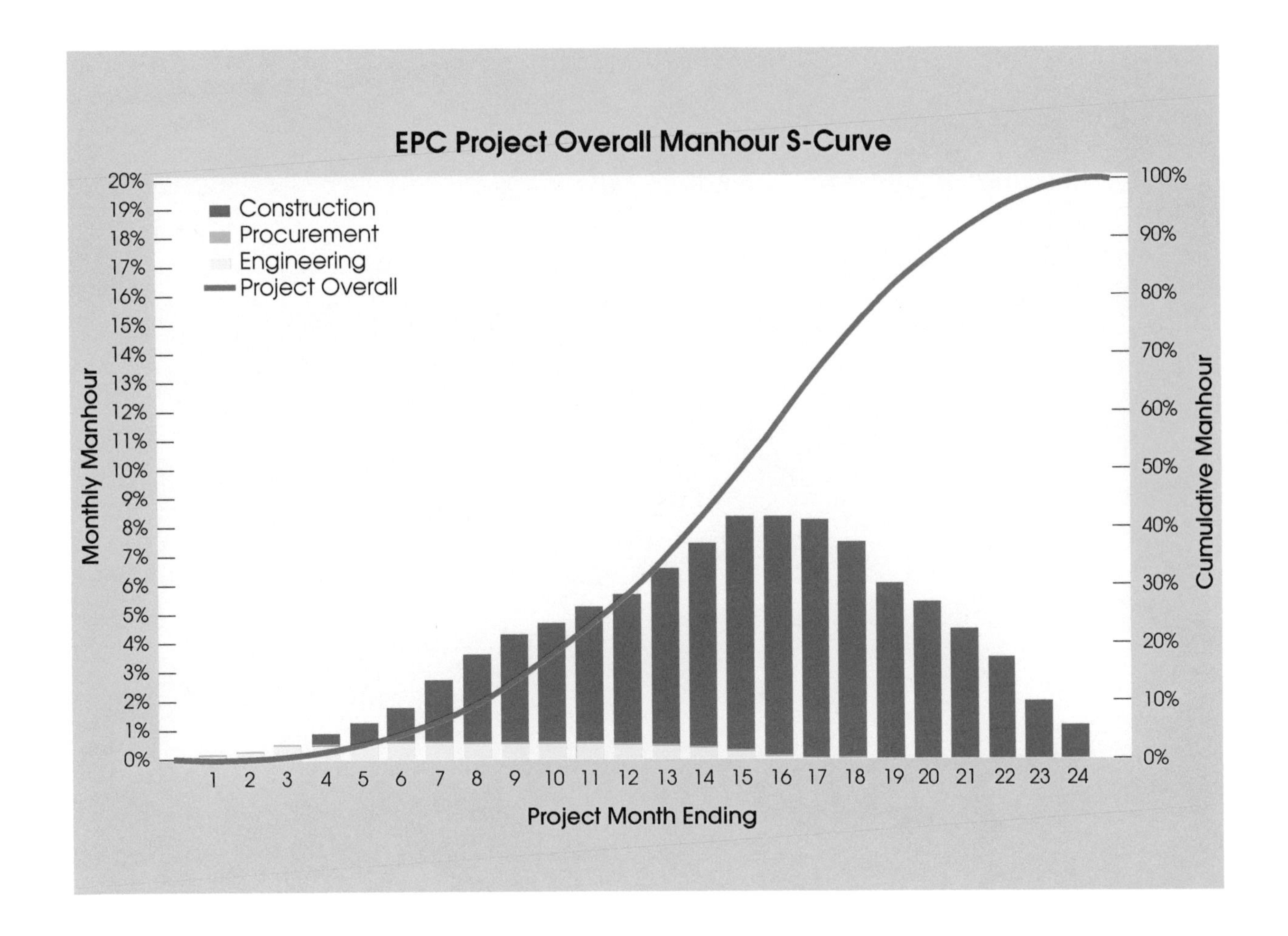
EPC Project Overall Manhour S-Curve
Construction
Procurement
Engineering
Project Overall
Monthly Manhour
Cumulative Manhour
Project Month Ending
20%
19%
18%
17%
16%
15%
14%
13%
12%
11%
10%
9%
8%
7%
6%
5%
4%
3%
2%
1%
0%
100%
90%
80%
70%
60%
50%
40%
30%
20%
10%
0%
1 2 3 4 5 6 7 8 9 10 11 12 13 14 15 16 17 18 19 20 21 22 23 24

Chapter 5

Cost Estimation

Cost estimation is the process by which the cost of doing something is assessed before it is completed.

So, are estimates just good guesses? The answer is no. There are a myriad of methods, mathematical models, scientific principles and engineering techniques which can be applied to the simplest level of cost detail to give the best chance of arriving at as accurate an estimate as possible. But the very word estimate implies that there will be a level of uncertainty involved.

Besides calculating the costs of the actual work to be done the uncertainty has also to be estimated and allowed for.

There is a difference between a cost estimate and a cost forecast. When talking of project costs these often blur together, but generally a cost estimate is what you have before you commence building your project and a cost forecast is what you have once you begin constructing your project. Or, a cost estimate is what you use to decide to commence the project and a cost forecast is what you use to control the project during execution.

Projects need to be planned and managed within scope, time, and cost. These elements are interrelated. Change one and you affect the others. For example, increase the scope and you will increase the cost and time. Perhaps time is the major constraint in the execution of the project so therefore the need to complete the project quickly will result in working overtime or nightshifts which will increase costs. A further issue to be considered in developing any cost estimate is the execution plan. That is the way the project will be achieved. In particular the contracting strategy will affect the estimate. There needs to be specific attention paid to allowing for the risks involved with the different ways that a type of contract can generate costs not directly related to the volume of work to be performed.

In order to assist project owners and managers to make decisions on whether to proceed with projects accurate estimates are required. From these estimates, rules of thumb can also be generated to help evaluate future estimates or allow project managers, engineers, to check the estimates they review or need to generate.

Generally in projects, three types of cost estimations are carried out. If you scan through all the available literature and information on the Internet you will read that there are 2,3,4, or even more types of estimate but we reckon through our experience that three is the number.

Preliminary, scoping or screening estimate
Detailed
Final

There are several ways to estimate each component of a cost estimate. The below methods are popular and widely used.

1. Costs from vendors
2. Costs from previous similar projects
3. Deriving costs from empirical formulas and the cost of raw materials

Estimate Accuracy

It may seem entirely obvious but the more information that is available to the estimator the more accurate the estimate will be.

This accuracy needs to be quantified because it is one of the components of the estimate, it is the metric that will determine part of the contingency. Below is a graphical representation of the way the accuracy is affected by the stage at which the estimate is performed. Most organisations have a corporate standard version of this or similar which is used by their estimators to develop the contingency.

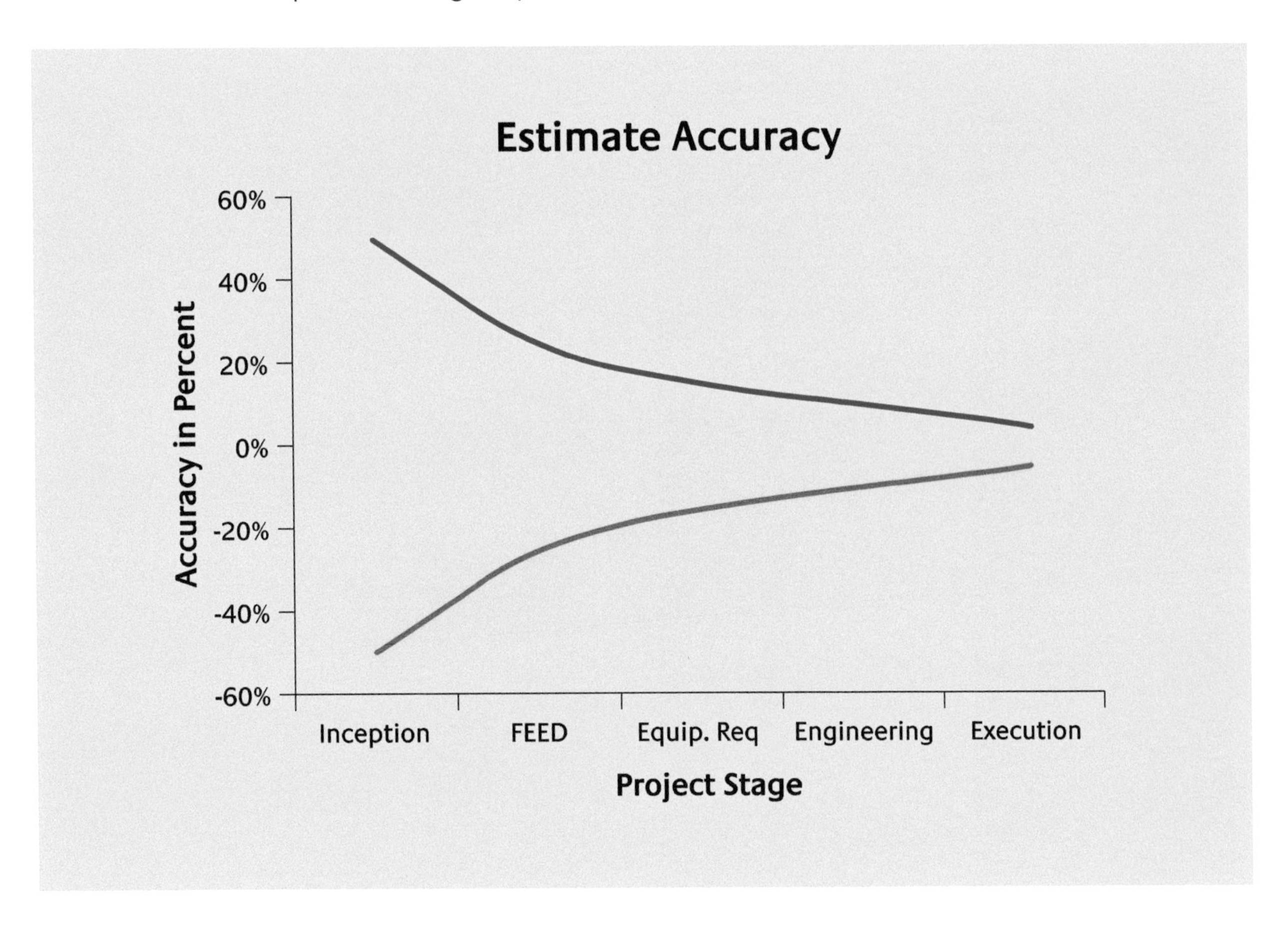

Why Students to Know about Cost estimation?

Students need to have cost awareness. The cost estimation of a project or a work will not only make students aware of cost components but also make them cost conscious. Any engineering graduate is expected to perform design calculation. However, Project engineering education will make students aware of design optimisations.

VALUE ENGINEERING is one of the most important concept of project engineering and that originates from the essence of cost estimation.

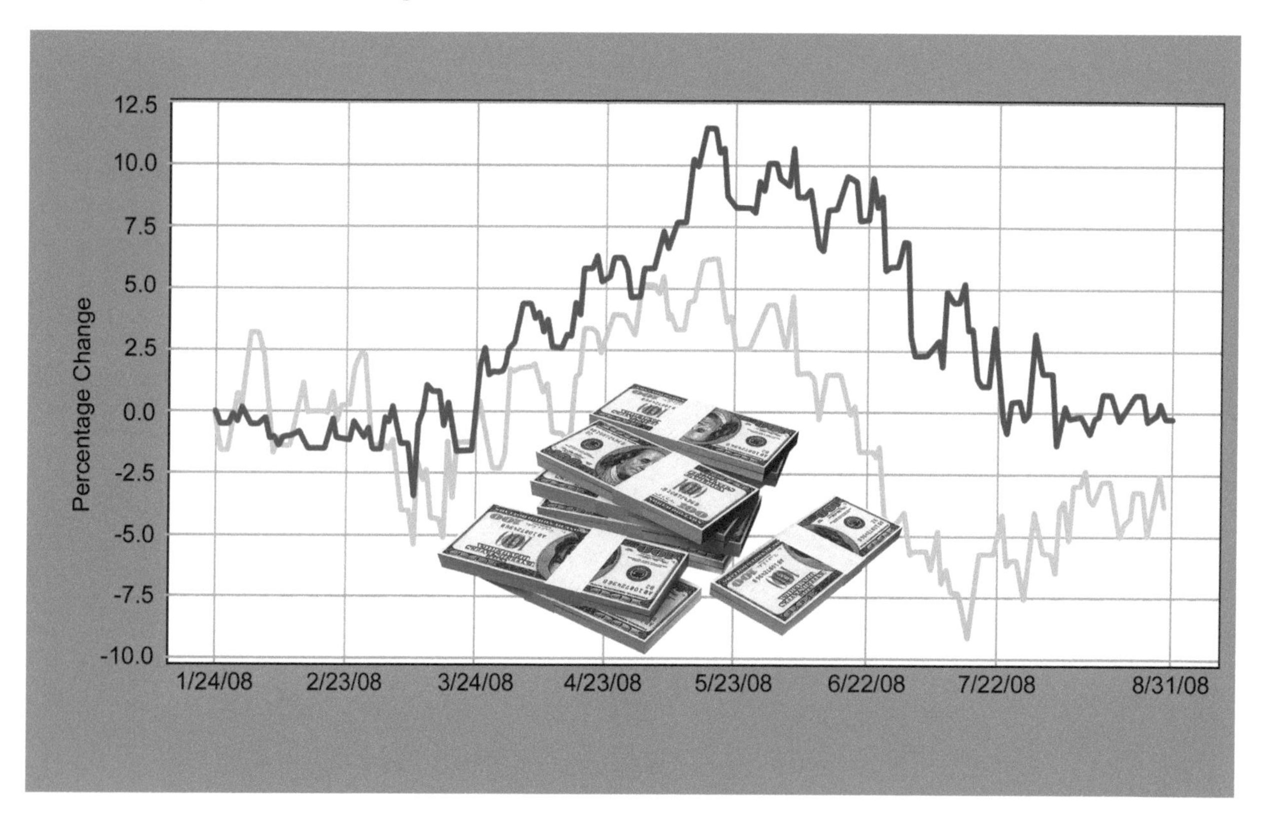

ITEM	Quantity	UNIT COST US $	Total Estimated Cost US $
PIPING ESTIMATED COST FOR THE FOLLOWING			
P-123 A/B NEW BIGGER SIZE PUMP	2	12391	24781
P-456 C ADDITIONAL PUMP	1	5489	5489
P-789 C ADDITIONAL PUMP	1	4535	4535
P-147 A/B PUMP IMPELLER / BASE FRAME CHANGE	2	1000	2000
P-148 A/B PUMP IMPELLER / BASE FRAME CHANGE	2	8728	17455
P-245 IMPELLER BASE FRAME CHANGES	1	2000	2000
E-234 ADDITIONAL KERO EXCHANGER	1	175142	175142
Material cost of steam tracing, painting and insulation = 5% of total piping Material Cost			11570
SUBTOTAL PIPING + INSULATION & PAINTING			**242972**
INSTRUMENTATION			
123 XV 12/34 (10" ALLOY STEEL VALVE)	2	10000	20000
134 PV 45 / 56 (12" CS CV)	2	9000	18000
456 TV - 345 (4" CS CV)	1	2000	2000
Instrument Bulk Material cost = 2.5% of Tagged item cost			1000
SUBTOTAL INSTRUMENTS			**41000**
SUBTOTAL INSTRUMENT + PIPING ====>			**283972**
Additional 25% TOWARDS TAX, DUTIES AND TRANSPORTATION etc.			70993
MATERIAL COST			**354965**
EXPEDITING/INSPECTION COST (5% OF TOTAL MATERIAL COST)			17748
MATERIAL + INSPECTION COST			**372713**
CIVIL			
CIVIL SUPPORTS = 5% OF PIPING COST			12149
ADDITIONAL FOUNDATION ETC IN M3	50	150	7500
ENGINEERING COST			**19618**
(5% OF TOTAL COST)			
MATERIAL + INSPECTION + ENGINEERING COST			**411980**
EXECUTION			123594
30% OD TOTAL MATERIAL COST			
GRAND TOTAL			535574
TOTAL COST		**SAY**	**5,40,000 US $**

Exclusions: 1. Pump + Impeller + Motor

Chapter 6

Project Interfaces

During project execution, the following major interfaces are dealt with:

1. Stake holder
2. Client
3. Supplier
4. User
5. Contractors
6. Regulatory Agencies
7. Political and Government

Stake holders - There are several definitions of Stake Holders. The common meaning is, stake holders are person or group inside the Company organization, who are accountable for the project. Some time Stake holders mean investor or investing organization.

Project team need to have regular interaction with the stake holders.

Clients are individual or corporate, for whom the project is being built

Suppliers supply project materials. It may be equipment, instrument and or several other items.

User - Project is built based on User's demand. User are group of people or corporate, who consume/use the project end product. User may a group of the same organization, who will in turn operate the plant upon project completion. These projects are often called CAPEX project.

Contractors are vendors, who are responsible for engineering, construction, commissioning and/or procurement. There may be one or more number of Contractors engaged in a project. There should be regular engagement of project management with contractors to keep project running on schedule, budget and quality.

Regulatory agencies are Public or government agencies, responsible for exercising authority over one or more subjects. Examples are Environmental Protection Agency (EPA), International labour organization (ILO), world health organization (WHO) etc. Project need to take care of these agency's agenda and rules. They hold capacity to cancel a project or stop production of a running plant.

Political and Government - Change of government policy could stop a project or make a project commercially unviable. Government officials need to engage time to time during project review, to make them feel responsible for the success of the project.

Out of the above all interfaces, the most interesting interface for a engineering student may be the contractor interface. However, it always depends on the individual. Interface engineering is part of detailed engineering of a plant where two or more EPC contractors are involved. It deals with the exchange of design, schedule and construction details between two or more contracting parties.

In good old days, one big plant used to be built by one EPC contractor. Now-a-days, installations are being put up in much bigger scale to make it more cost effective. The time frame from when the investment decision has been made to commissioning is also getting shorter. Mainly to avoid any schedule or budget overrun risk, the owner may divide the whole complex into several smaller contracts implemented by multiple EPC contractors. The timely and accurate exchange of design, schedule and construction information between design offices becomes important. This exchange of information is to minimize rework at a later date and avoid last moment surprises during commissioning.

It becomes the task of project management team to act in advance to identify the interfaces before the packages go for bidding. In case, the interfaces are not identified or not identified properly, then the detailed engineering contractors need to depend heavily on accuracy of FEED (Front end engineering design). Design development/changes are normal between FEED and detailed engineering. These changes may not be captured in the integrated design until unless detailed engineering design information are exchanged between two or more parties at contract boundaries, consolidated and subsequently re-confirmed. Shortcoming in interface identification typically leads to inadequate flare design, existence of same line in two different sizes in two contract boundaries, missing lines, mismatch of pipe elevations at expected match lines, clashes of one contractor's cable trench clashes with other contractors under ground pipe etc. Mostly, it is the utility systems and product/raw material systems that get adversely affected by inadequate or no exchange of design information.

Identification of interfaces is a critical task. It needs involvement of all engineering disciplines to identify their respective discipline interfaces between different unit/ project boundaries. Construction engineers also need to be involved to identify the construction interfaces, i.e welding and hydro test sequence, schedule between two contract boundaries etc.

During interface identification period, there is always a probability to miss an interface. The detailed engineering contractor should recognize these potential shortcomings early during their detailing engineering phase.

Effective early identification of interfaces also helps to reduced increased scope which may occur if interfaces are not adequately recognized.

Identification of interfaces is normally being done by FEED contractor or by an independent Project management team. As this identification work can be executed under one roof, it may be termed as manageable. However to make agreements between two unwilling EPC contractors, working from two different geographical locations, requires a completely different approach. This becomes an interface engineer's skill and expertise to resolve the issue strictly under project guideline without any extra cost. The responsible person from the owner side requires strong leadership quality and drive and good co-ordination skill to manage changes arising from interface.

In interface engineering, the first step is to identify interfaces. The second step is to get the parties to agree upon in content and schedule of information exchange. The third step is to ensure exchange of proper design information occurs in the time frame agreed. Finally and the most important task is to ensure the interfacing parties agree upon each other's information.

Chapter 7

Assessment and Mitigation

The uncertainties and threats of a project are collectively defined as project risk. Risk is a measure of uncertainties. An uncertainty comes from an absence of information or lack of understanding of the outcome of a decision. Risk is a negative phenomenon, i.e. risk is the event or cause that drives the project out of its desired trajectory.

Every project has to undergo some form of risk. Project organization has to measure, take a proactive approach towards risk mitigation and management. To become successful in project execution an organization has to be honest in identifying project risk. A collective decision has to be taken by the project organization and investors together whether to go ahead with the project execution. To go ahead the management has to accept some risk and to the manage them till the end of the project.

In the EPC concept, from a lump-sum contractor's point of view, every risk is some amount of money and additional cost. Any EPC contractor has to mitigate the risk at project cost but from an owner's point of view, many project organizations may deny this fact and may say that all Risk may not be formulated in terms of money. As for example, the risk may be of reputation, however every project is a business and business is nothing but a deal in terms of money.

Type of Project Risk

In a very simple manner, project risks can be categorised as below.

1. Organizational
2. Project Scope
3. Regulatory Rules
4. Market
5. Customer
6. Subcontract & Suppliers
7. Budget
8. Schedule
9. Funding
10. Resources

Quantifying Project Risk

The Planning department of a project organization should to hold a planning meeting to develop a risk management plan. In A simple way to describe these activities are:

- Identifying risk
- Quantifying the risk
- Adopt suitable risk management plan against each risk

In the beginning of any project, a risk management team is formed, that includes the project manager, selected project team members, Owner and possibly the maybe investor(s) (In high level risk meeting). One person is identified among the project organization with the responsibility to manage the risk planning and execution activities.

At the beginning stage of the project during the above said noted risk meetings, the following items for the project are discussed and agreed upon among the team members.

Template of Levels of Risk

Template of Probability

Template of Impact

Probability & Impact Matrix

Two types of risk analysis are normally performed.

1. Qualitative Risk Analysis

2. Quantitative Risk Analysis

These two widely known and well written methods are described below just to give the reader an impression of what are these analyses stands for.

Qualitative Analysis

This is a process of determining project risk on cost, schedule, scope and quality for further analysis of relative probability of occurrence and impact.

The project risk register gets updated after the qualitative risk analysis. As mentioned above, this analysis is performed based on approved project risk management procedure, using probability and impact matrix. The project risk committee evaluates the risk based on risk category and take necessary countermeasure based on impact assessment.

The project risk committee should constitute of senior and experienced project members who has have the necessary experience and attitude towards (acceptance of certain) risk gained from similar projects. There should be an agreement in a project committee on the impact and occurrence of certain risk.

The qualitative risk assessment provides ground work towards Qualitative risk analysis.

The most common approach to this process is to create a risk rating matrix. Here's a quick sample of a risk rating matrix of a EPC project:

Risk	Impact	Probability	Score
Engineering	Very high	Medium	High
Technology	High	Medium	High
Schedule	Medium	Medium	Medium
Vendor	High	Medium	Low

Within a project, the risk committee has to determine which of these risks deserve additional analysis. Typically the risks with a medium score or higher should be taken seriously and are promoted to quantitative risk analysis.

Quantitative Analysis

Quantitative risk analysis is a process of quantifying project risk in a numerical way. Qualitative risk analysis identifies project risk and quantitative risk analysis work further on those identified risk and give it a numerical identification of its impact on overall project.

Input to quantitative risk analysis is as mentioned before, project risk management plan (Schedule, cost, quality & safety management plan etc), risk register, scope document (new unproven technology or copy of another project) , if similar project then the previous project risk database , etc.

After quantitative risk analysis the project risk register gets updated.

To do the quantitative risk analysis, a project need database on probability of occurrence of a particular risk. Risk again may be of cost, schedule or any other as mentioned in chapter 2.

There is no shortcut for preparation of a database, this database can not be purchased or sold. Either it may be obtained from previous similar project or It has to be made for a particular project. The More data gathered the more chance to get near to an accurate forecasting of a risk.

Chapter 8
Construction

Construction is a process of building pre determined facilities. Construction in any project is managed by Construction manager. The construction organization structure depends on type of project. The organization chart of a construction department of any EPC project of process unit (Industrial construction) is described in the following chart.

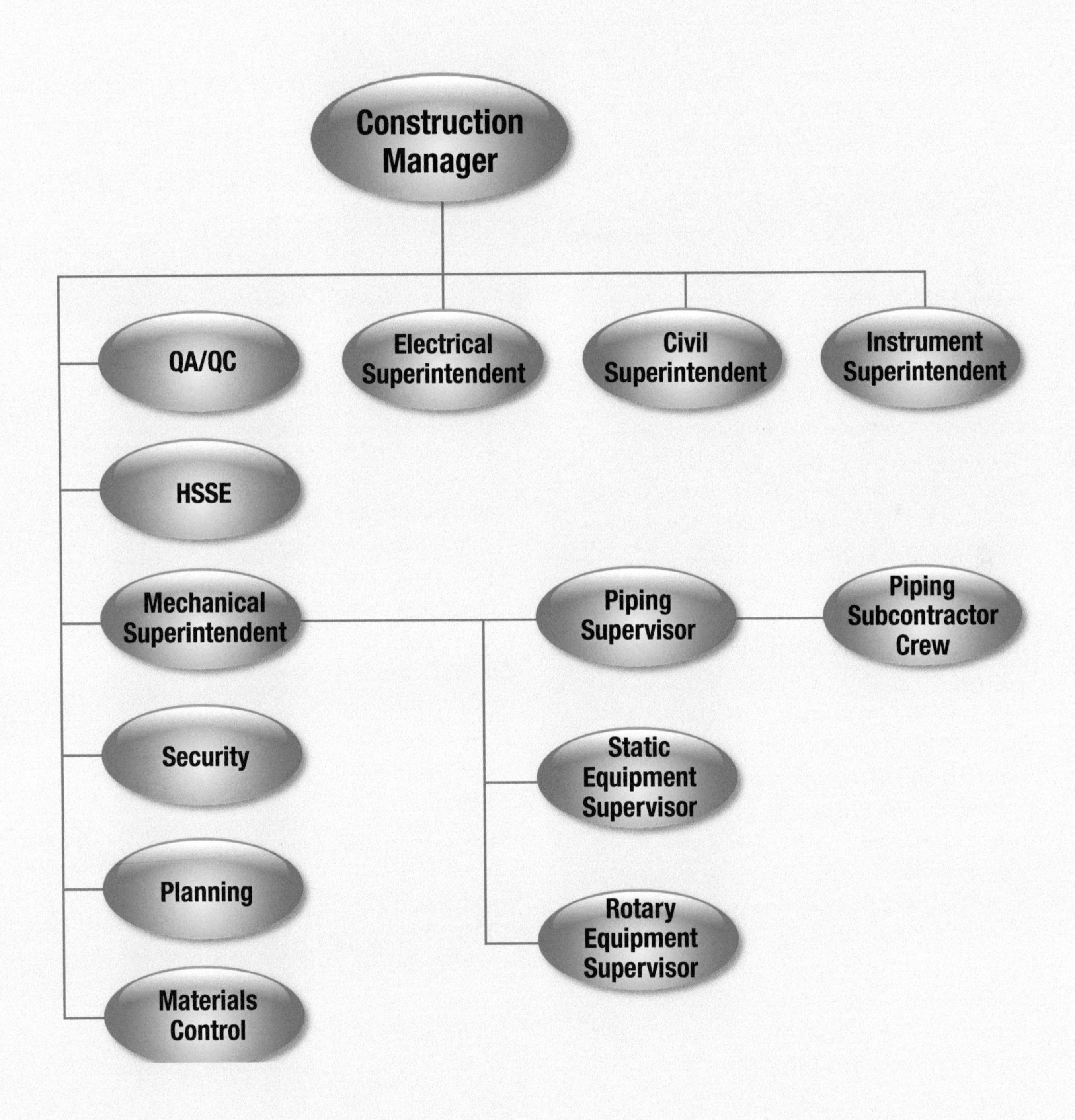
Construction Manager
QA/QC
Electrical Superintendent
Civil Superintendent
Instrument Superintendent
HSSE
Mechanical Superintendent
Piping Supervisor
Piping Subcontractor Crew
Static Equipment Supervisor
Rotary Equipment Supervisor
Security
Planning
Materials Control

The construction project varies widely on type of project. Like , infra structure projects (Road, Dam, Bridge etc.) constitute more civil work and superintendant and staff members come from civil back ground. In case of electrification project, construction manager comes from electrical back ground and the project organization varies accordingly.

In case of lump sum project, if work is divided between subcontractors, the construction organization is dominated by Supervisors. If the same work is being done by the same organization, (without sub-contracting), the construction organization need to employ much more staff and crew members of different disciplines

Apart from progress, Safety and Quality are two most important aspects of Construction management. In every construction project, depending on project criticality, project employs Health Safety Environment (HSE) specialists. HSE group enforce project laid down safety rules, conduct daily awareness meeting with workers. Construction quality manager (Quality assurance and Quality control) is expected to be well conversant with construction related codes & standards. Quality control department reports to construction manager and enforces project quality policy to facility construction.

Man, Machine and materials are three most important aspects of any construction project.

Man

In a mega project, where there is a requirement of large amount of manpower, construction management mobilises manpower from different region/different part of globe. This is done to offset the effect of regional holiday/festive seasons. Accommodation of labour force at site or away from site becomes priority of construction management. This labour accommodation is occasionally called CAMP. Health and safety of the camp with minimum facilities (varies from project to project) depends on local government labour law or basic rules of the contracting company.

Machine

Day by day, the construction is becoming more and more machine intensive. The picture of manual excavation of a big plot is becoming a story of the past. When construction contractors mobilise machineries , it needs workshop and space for maintenance and inventory. Project health and safety rules are applicable to these workshops and lay down areas. Especially cranes and heavy machineries need lots of space for overhauling activities. Identification of lay down area in a plot is essential to reserve space for this activity. Authorised personnel inspect machines periodically and certify for fitness.

Material

Materials are sourced much before erection and need space to preserve and store. In case of offshore project, where plants are often built in a modular fashion, materials are not allowed to keep at construction site for longer period. In case of on shore project, large lay down areas are used to store electrical, instrument, mechanical or piping items.

To accelerate site piping erection work and minimise site welding, construction management may decide to install pipe spool fabrication shop. Where spools are fabricated using automated welding machines in a controlled environment.

Similar fabrication shop is common for insulation and painting.

Depending on client HSSE policy, construction contractor construct grit blasting and/or painting shop.

In case of civil construction, construction contractors need concrete. Depending of size of project and frequency of requirement, construction contractor either install cement batching plant or sub contract cement mixture. (Popular cement mixing subcontractor is Ready Mix. These contractors can supply different grades of concrete mounted on a truck. The cement need to be used with in a stipulated time frame.

Construction projects can be categorised as under:

1. **Building construction**
2. **Infrastructure construction (Rail, road, bridge etc.)**
3. **Industrial construction**

Building Construction

Building construction group are engaged to construct small to tall high rise buildings. These days, house/apartment buildings are made up of pre fabricated building materials. However, conventional building construction with wood, mortar, bricks and concrete are also equally popular. In case of conventional way of building of houses, contractors build scaffolding around the building to facilitate labour safety and safe access. In high rise building, in addition of scaffolding, use of Tower Crane and elevators has become very common practice. In big cities, we often see high rise under construction buildings with tall cranes in the horizon.

Standard types of personnel, necessary to construct a building is as below:

1. Construction manager
2. Contract manager
3. Architect
4. Construction superintendent
5. Planning engineer
6. Personnel manager and HR team
7. HSE officer and HSE team
8. Material Manager and it's warehouse logistics team
9. Quality assurance and Quality control officer and it's team
10. Mason and it's team
11. Carpenters
12. Plumbers
13. painters
14. HVAC (Air conditioning) contractor
15. Electrical crew or contractor
16. Crane and it's crew
17. Truck/dumper and operator/Driver
18. Plumber, Pipe fitter, welder and associated men and machine.
19. Security manager and it's team
20. Paramedic

Infrastructure Construction (Rail, Road, Bridge etc.)

Team of infrastructure construction is same as building construction. However, Infra group is more experienced in heavy rigging and construction. e.g. during bridge construction, piling under water or ground is essential to maintain foundation elevation. Piling is a much specialised work and crew are expensive.

During infrastructure construction work, keeping constant touch with Government official becomes most important as project need to obtain day to day permit to work from government body. Normally senior and experienced staffs take care of government relation activity. Government relation officer liaise with Government official for Labour induction and other labour related activities.

Industrial Construction

Industrial construction is becoming more and more mechanised day by day. Construction contractor need to maintain workshop to maintain its heavy construction machineries. Number of crew working in a project depends on the project size e.g. it can vary from 1,000 to 60,000 and even higher.

Construction work can be categorised as below:

1. Civil construction - Concrete
2. Civil construction - Structural Steel fabrication & erection
3. Mechanical construction - Rigging or Erection
4. Mechanical construction - Piping
5. Electrical and instrument construction - Erection of tagged items, laying of cables etc.

Below are description of each category of construction work and it's popular thumb rules.

1. Civil Construction Work

Deals with concrete. Concrete is a mixture of sand, cement, cementitious materials (Fly ash, lime stone, granite etc) and water. Concrete solidifies and gets hard after mixing with water and placement due to a chemical process known as hydration. The water reacts with the cement, which bonds the other components together, eventually creating a robust stone-like material. Different ratios of mixture make its different grade of concrete.

Sequence of standard civil construction work is as below:

a. Making basic foundation form work with wooden pallets/metal sheets.
b. Filling the hollow foundation block with structural steel rods as per design.
c. Pour concrete in the foundation block. Concrete should be as per quality standards mentioned in the project specification.
d. Curing - Water keep the concrete wet for a duration, as mentioned in the project specification.

Below items are to be mentioned as they cover important part of civil construction.

Concrete Batching Plant – Is a facility, where different ingredients of concrete are mixed in a defined ratio and transported to construction site. These concrete plants are called as batching plant, because hoppers are loaded in batches with specified amount of raw materials of concrete and they can change the grade of concrete in batches, if desired. Ready Mix is a popular brand of concrete used in all type of civil constructions.

Concrete mixture – A devise, mounted on a wheel or mounted on a truck that homogeneously mixes concrete, sand and water. Concrete mixture truck is to transport concrete mixture from batching plant to the construction site.

Concrete Pump – Concrete pump is a devise that pumps concrete from grade to a higher elevation.

2. Civil Construction Structural Steel Fabrication and erection

– Involves cutting, grinding, welding and erection of steel work.

Structural steel is manufactured in various different shapes e.g. Beam, Channel, angle, plate and hollow sections. These section's dimensions' are based on different standards.

Popular standards structural steels commonly used in industry are as below.

1. Steels manufacturer in Europe follows European standard – e.g. EN 8, S195J2. S denotes structural grade steel , while E denoted engineering grade steel. First three numbers denotes yield strength. The most commonly available yield strength grades are 195, 235, 275, 355, 420, and 460. Higher grades are also available on demand.

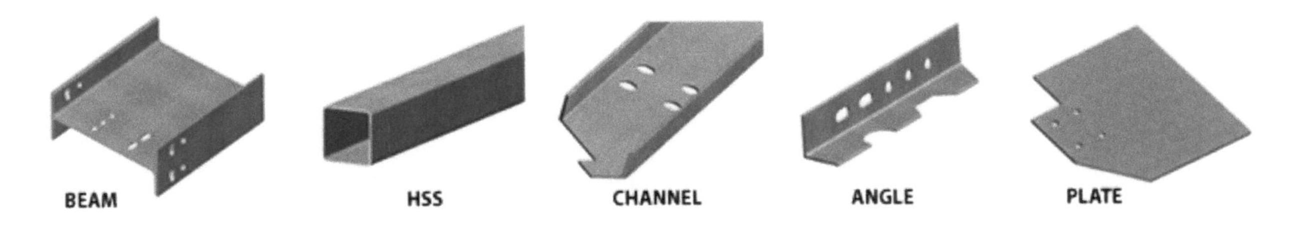

2. Steels used for building construction in US are standard alloys, identified and specified in ASTM standards. All ASTM numbers starts with A, e.g.

   Carbon steels: A36 – structural shapes A529 – structural shapes and plate

   High strength low alloy steels: structural shapes and plates – A270, A441, A572, A618, A992

   Corrosion resistant high strength low alloy steels: structural shapes and plates – A242, A588

   Quenched and tempered alloy steels: A514, Structural shapes and plates, A517, Boiler and Pressure vessels.

Points to remember that steel starts melting at 1130 degC and Pure Steel with 0% carbon starts melting at 1492 degC. Steel with more than 2.1% Carbon is called as cast iron.

Equipments and methods used in Steel construction

1. **Types of Welding Processes**
   Manual Shielded Metal Arc Welding (SMAW): Welding rod (electrode) placed in an welding holder and manually welded by a welder.

   Submerged Arc Welding – flux cored arc welding – (FCAW) (Automatic/Semi automatic):

In case of automatic flux cored arc welding, machines controls the wire feed and travel speed, where as in semi automatic welding machine, Hand held equipment controls the wire feed only , welder controls travel speed.

2. **Type of Consumable/Electrode:**
   a. SMAW – examples – E 6013, E-7018 etc.
      E = AWS electrode
      60 = Tensile strength (Ksi)
      X = Position (1= Flat, Horizontal, Vertical, Overhead), (2= Flat, Horizontal), (3= Flat)
      X = Coating type

   b. Standard Size of electrode (Max Outer diameter) = ¼", 3/16", 5/32"
   c. Oven is used to heat electrodes before used = 2-4 hrs at 450degF to 500 degF, then keep at 250degF until used.

3. **Welding Certification**

WPS – Welding procedure specification – it is documentation prepared for particular process of weld on a specific weld geometry and weld parameters (Based on an approved PQR).

PQR – Procedure Qualification record is a record of a test based on an approved WPS, to determine of validity of a specific set of welding parameters.

4. **Inspection**

Radiography: X Ray is passed through the weld and base metal to take the image on a photographic film. The common welding defects are slag, porosity, lack of fusion, lack of penetration.

Magnetic Particle Test: In this test, magnetic field is created over base metal. Iron power is sprayed over the magnetic field. In case of any defect, iron powder presence on the surface indicates discontinuity.

Ultrasonic Testing: High frequency sound waves are passed through weld and base metal. The sound waves are analyzed by special instrument to detect defect in the weld joint.

Dye penetrant testing: Dye is sprayed over the test surface. Dye penetrates the defective surface and gets dry in the air. There after, a developer liquid is sprayed over the surface again to identify the defective area.

3. Mechanical Erection/Rigging

These days rigging is mostly mechanized and different types of cranes are used for different applications. Below is a typical rigging study.

4. Mechanical Construction - Piping

Codes and Standards - ASME, ANSI, ASTM, AGA, API, AWWA, BS, ISO, DIN etc.

ASME B 31.1 - Pressure Piping

B31 is the code for pressure piping. This code is developed by AMERICAN SOCEITY OF MECHANICAL ENGINEERS This code covers Process Piping, Power Piping, Fuel Gas Piping, Pipeline Transportation Systems for Liquid Hydrocarbons and Other Liquids, Refrigeration Piping and Heat Transfer Components and Building Services Piping.

ASME B31 was earlier known as ANSI B31. Here below are the commonly used ASME standards

B31.1 - Power Piping

This code prescribes minimum requirements for the design, materials, fabrication, erection, test, and inspection of boiler external piping for power boilers, electric generation stations etc.

B31.3 - Process Piping

This Code covers rules for piping of refineries; chemical, pharmaceutical, cryogenic plants etc.

This Code specifies requirements for materials and components, design, fabrication, assembly, erection, examination, inspection, and testing of piping.

B31.4 - Pipeline Transportation Systems for Liquid Hydrocarbons and Other Liquids

This Code defines rules for the design, materials, construction, assembly, inspection, and testing of piping transporting liquids such as crude oil, condensate, gasoline etc. This code covers piping consists of pipe, flanges, bolting, gaskets, valves, fittings etc. It also includes hangers and different type pipe supports.

B31.5 - Refrigeration Piping and Heat Transfer Components

This Code defined rules for the materials, design, fabrication, assembly, erection, test, and inspection of refrigerant, heat transfer components

B31.8 - Gas Transmission and Distribution Piping Systems

This Code defines the rules of the design, fabrication, installation, inspection, and testing of pipeline facilities used for the transportation of gas. This Code also covers safety aspects of the operation and maintenance of those facilities.

During piping section, piping contractors use Crane and other conventional lifting tools, rollers etc.

5. Electrical & Instrumentation Construction

Electrical and instrumentation construction work mainly concentrated around erection of tagged items (Motor , Transformer etc), Cable laying (under ground & above ground), cable tray erection etc. During electrical cable pulling, use of roller and winch has become increasingly common. For erection of Transformers, motors and other tagged items, electrical contractors make use of crane and other common erection tools and tacles similar as mechanical group.

Standard co**des applicable for Electrical and Instrumentation**

NPS	DN mm	OD in (mm)	Wall thickness (in (mm)) SCH 5	SCH 10s/10	SCH 30	SCH 40s/40 /STD	SCH 80s/80 /XS	SCH 120	SCH 160	XXS
1/8	6	0.405 (10.29)	0.035 (0.889)	0.049 (1.245)	0.057 (1.448)	0.068 (1.727)	0.095 (2.413)	—	—	—
¼	8	0.540 (13.72)	0.049 (1.245)	0.065 (1.651)	0.073 (1.854)	0.088 (2.235)	0.119 (3.023)	—	—	—
3/8	10	0.675 (17.15)	0.049 (1.245)	0.065 (1.651)	0.073 (1.854)	0.091 (2.311)	0.126 (3.200)	—	—	—
½	15	0.840 (21.34)	0.065 (1.651)	0.083 (2.108)	—	0.109 (2.769)	0.147 (3.734)	—	—	0.294 (7.468)
¾	20	1.050 (26.67)	0.065 (1.651)	0.083 (2.108)	—	0.113 (2.870)	0.154 (3.912)	—	—	0.308 (7.823)
1	25	1.315 (33.40)	0.065 (1.651)	0.109 (2.769)	—	0.133 (3.378)	0.179 (4.547)	—	—	0.358 (9.093)
1¼	32	1.660 (42.16)	0.065 (1.651)	0.109 (2.769)	0.117 (2.972)	0.140 (3.556)	0.191 (4.851)	—	—	0.382 (9.703)
1½	40	1.900 (48.26)	0.065 (1.651)	0.109 (2.769)	0.125 (3.175)	0.145 (3.683)	0.200 (5.080)	—	—	0.400 (10.160)
2	50	2.375 (60.33)	0.065 (1.651)	0.109 (2.769)	0.125 (3.175)	0.154 (3.912)	0.218 (5.537)	0.250 (6.350)	0.343 (8.712)	0.436 (11.074)
2½	65	2.875 (73.02)	0.083 (2.108)	0.120 (3.048)	0.188 (4.775)	0.203 (5.156)	0.276 (7.010)	0.300 (7.620)	0.375 (9.525)	0.552 (14.021)
3	80	3.500 (88.90)	0.083 (2.108)	0.120 (3.048)	0.188 (4.775)	0.216 (5.486)	0.300 (7.620)	0.350 (8.890)	0.438 (11.125)	0.600 (15.240)
3½	90	4.000 (101.60)	0.083 (2.108)	0.120 (3.048)	0.188 (4.775)	0.226 (5.740)	0.318 (8.077)	—	—	0.636 (16.15)

Chapter 9

Pre-commissioning & Commissioning

What is Pre-commissioning?

Pre commissioning activities include the preparation for commissioning or testing the plant or any part prior to be ready for commissioning. Example of the pre-commissioning activities may include checking to confirm containment, continuity, all materials are present, internally clean, free of foreign materials etc.

Typical activities of pre-commissioning activities are flushing and blowing activities, initial leak/tightness testing etc.

What is Commissioning?

Commissioning means testing and inspection to ensure plant is in full working order to the specified requirements before the plant is RFSU (Ready for Start Up).

Commissioning is the Quality assurance process that includes scheduling and meticulous documentation of several tests that is carried out in a safe manner to confirm the facility is built and ready for operation as per project specification.

Commissioning activities include the preparation for operation, e.g. dynamic function testing, simulation runs (with benign fluids e.g. air/water/nitrogen).

Mostly in pre-commissioning & commissioning team, Commissioning Manager leads all activities. In the following paragraph, the roles and responsibilities of a Commissioning manager are described to give a general understanding, what are the functions of a commissioning team.

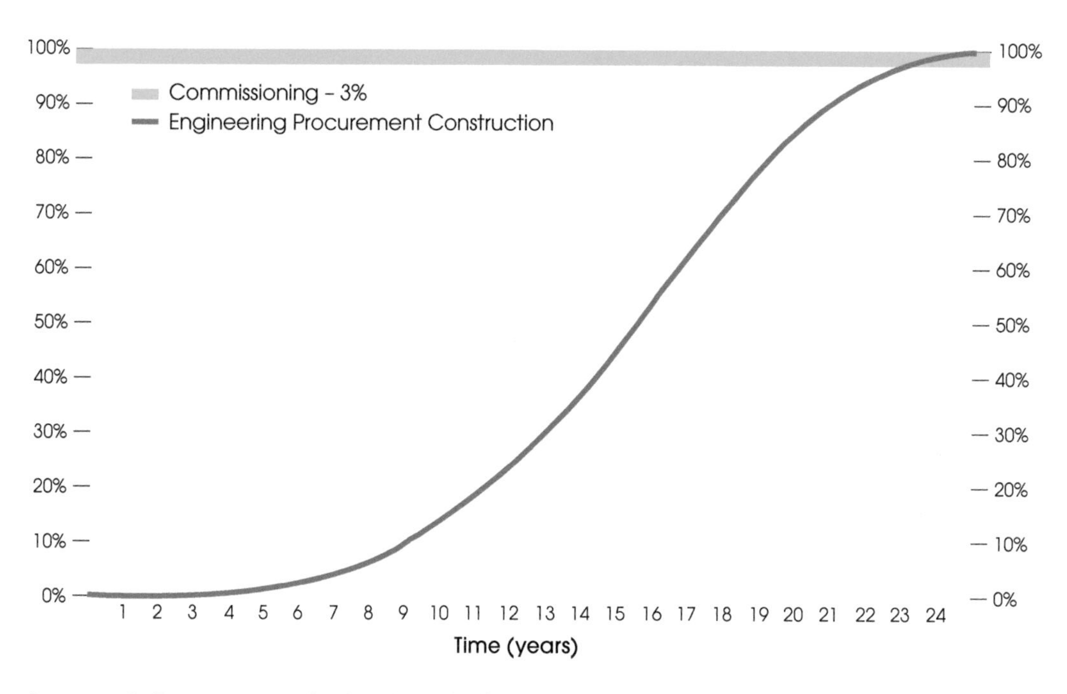

Percentage of Pre-commissioning & Commissioning Activities in a EPCM Project

From the above pictorial, pre-commissioning (Pre-com)/Commissioning (Comm) is represented as approximate 3% of the total overall project progress. A typical breakdown of the man hour represented by the 3% figure is 60% = Pre-comm; 40 % = Comm

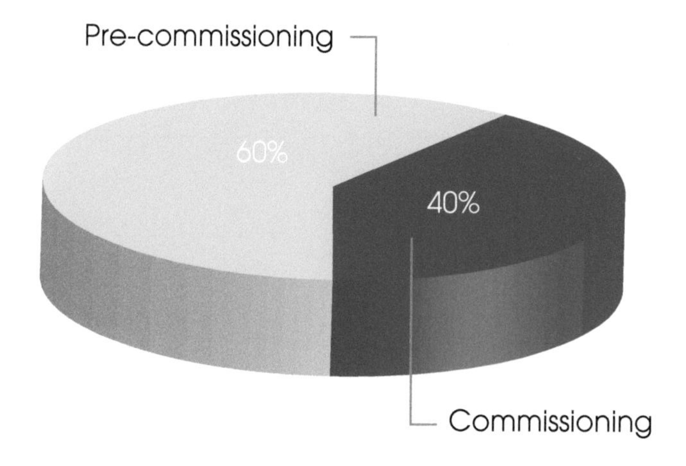

What are the roles and responsibilities of a Commissioning Manager?

The roles and responsibilities of a typical commissioning manager have been described here to give an impression about the function of pre commissioning & commissioning group.

These roles indeed varies from project to project depending on project type e.g. Contractor/ Project Management consultancy/Client and nature of project e.g. Process plant , mechanical plant etc.

Home Office Work

1. Participating in design reviews & 3D model reviews, ensuring compliance with project specification that relates to commissioning.
2. Taking a lead role to develop the systemisation definition and maximise early testing opportunities
3. To ensure that cleanliness and testing are fully incorporated into commissioning plans and procedures and method statements.
4. Preparation and review of various commissioning document deliverables e.g. commissioning execution plan, pre commissioning method statements, commissioning manuals, high level operating manuals, operating procedure/ work instructions etc.
5. Review and develop the commissioning schedule and identify all external and internal interfaces.
6. Carries out a detailed assessment of the work involved in the individual group as well as discipline with respect to Pre-Commissioning and Commissioning.
7. Preparation of manpower requirements for pre-commissioning, commissioning and start-up and prepare organization chart. Commissioning manager recruits , assess the available resources and possible constraints in the execution .

8. Define the system/sub-system definition.

9. Development of pre-commissioning, commissioning and start-up logic sequences used to prepare the high level schedule. Commissioning manager coordinates with the pre-commissioning and commissioning team and ensures the development of integrated plans including the scope of pre-commissioning and commissioning activities, duration for each, scope of interlinked activities, resources required, constraints regarding the same, as well as availability of tools and machinery in addition to overseeing the preparation of pre-commissioning and commissioning schedules based on the quantum of job and duration for each activity as well as prioritize system completion requirements in order to steer construction completion.

10. Provides guidance to the pre-commissioning and commissioning team in analyzing the scope as per contract and battery limit/guarantees as well as ensure the preparation of procedures for systematic and sequential pre commissioning/commissioning/start up of plant.

11. Identification of needs and preparation of commissioning personnel, Vendor's/ Licensor's mobilization schedule.

Site Work

1. Participation in development of sub-contracts for specific operations like, pipeline commissioning, chemical cleaning, specialist technical labor supply.
2. Definition of resources, materials, consumables, and special tools required for pre-commissioning, commissioning and start-up.
3. Review of pre-commissioning and commissioning database.
4. Ensures that all safety requirements and procedures are followed during Pre-commissioning, commissioning, and start-up activities.
5. Oversees and ensures all construction activities are completed prior to taking up the system for commissioning and ensure the setting up of safety procedures for the same.
6. Active involvement in defining the construction completion priority as per the pre-commissioning, commissioning and start-up sequence logic.
7. Coordinates with other departments to resolve interface problems and hold regular review meetings with the client on issues such as HSE conformance and pre-commissioning and commissioning progress as well as meetings with the vendors to resolve any equipment related problems. Reviews reports generated by Commissioning Engineers and provide support and technical guidance as and when required.
8. Preparation of daily, weekly and monthly reports as required.
9. Management of all documents necessary for pre-commissioning, commissioning and start-up and final documentation handover.
10. Management of pre-commissioning and commissioning database and punch list.
11. Management of start-up of the facilities. Management of performance test run execution and provisional acceptance of facilities.
12. Initiating corrective actions and verifying their proper implementation.
13. Pre-commissioning and commissioning team hands over the plant to client in line with the commissioning start-up schedules and client requirements.
14. Preparation of final closeout report related to pre-commissioning, commissioning and start-up.

Commissioning/Pre-commissioning Manager's place in an organization

Commissioning manager assists Project manager in delivering pre-commissioning and commissioning scope of work in terms of Quality and Safety.

Typical activities during pre-commissioning which confirms system compliance to the Project Specification

1. Cleaning
2. Reinstatement
3. Loop checks
4. Motor Solo Run
5. Alignment
6. Final Box-up / Vessel loading
7. Dry function test
8. Low pressure lead test
9. Safety walkdown and punch listing of system with regard to system compliance to project specification and project HSE guidelines.

Typical activities during commissioning which confirms system compliance to the Project Specification

1. Fill the system with test medium and pressurise
2. Tightness test
3. Establish flow and circulate
4. Alignment
5. Dynamic Function test
6. Refractory dry out
7. Degreasing and rinsing Amine system (Post Dynamic Function test)
8. High pressure Tightness test
9. Nitrogen Purging / inerting
10. Integrate systems for start up
11. Chemical preparation and filling
12. Handover

The standard organization structure of a commissioning team is as under:

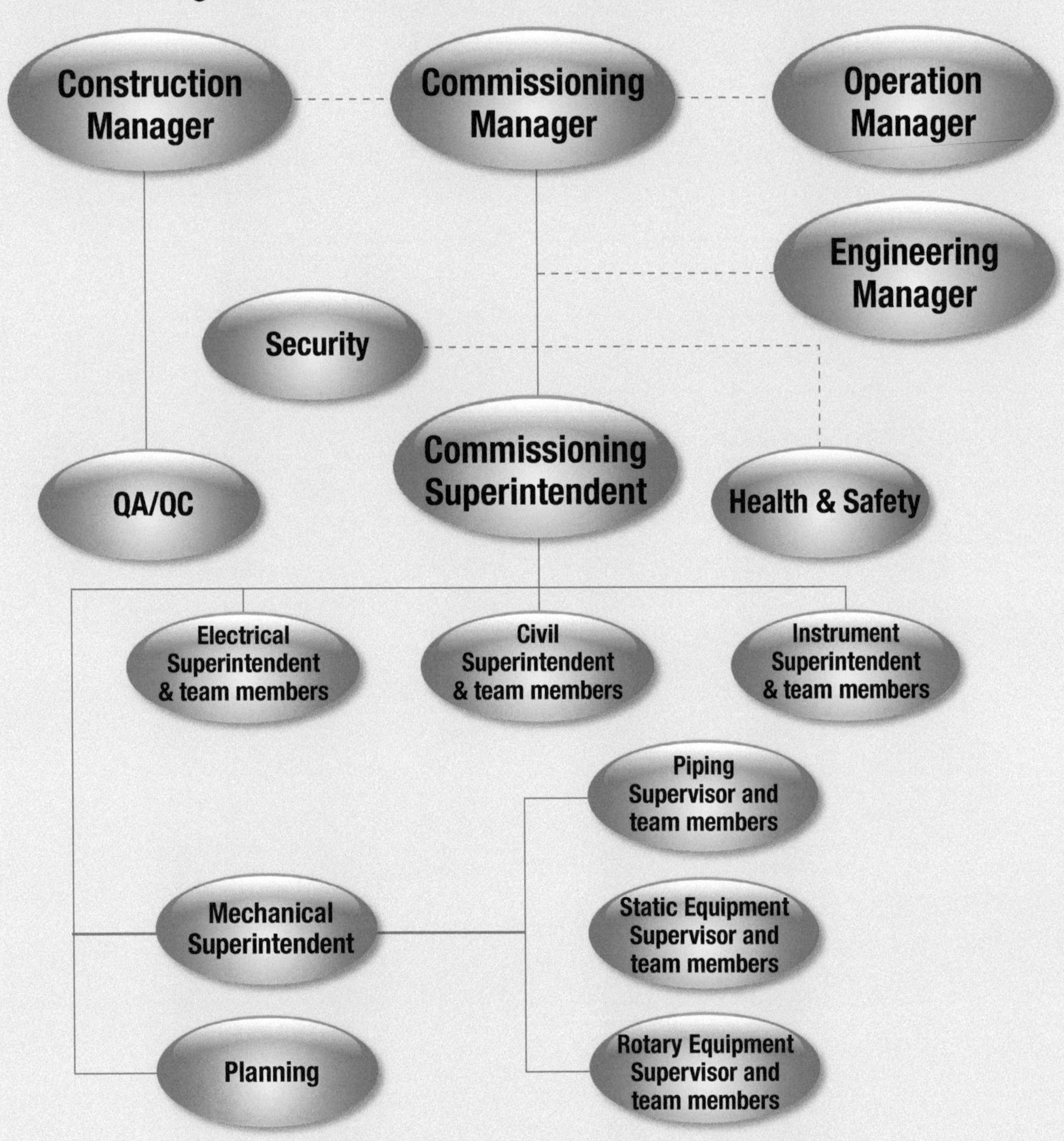

Reference

1. Multi-Discipline Project Engineering
   – S.Moulik, March 2010

2. Introduction to Cost Estimation
   – S.Moulik & Frank Feely – Feb 2011

3. How to avoid Interface Engineering problems?
   – S.Moulik & K.Saito – Hydrocarbon Processing – June, 2010

CPSIA information can be obtained
at www.ICGtesting.com
Printed in the USA
BVHW021948261118
534041BV00011B/116/P